Contents

1 Introduction to Intermodalism

Intermodal transport, at its core, is the seamless movement of goods using multiple modes of transportation—rail, road, sea, and air—without handling the cargo itself during transfers between modes. This integrated approach offers significant advantages over single-mode transport. Efficiency gains arise from optimized routes and reduced handling times, leading to faster delivery and lower overall costs. Furthermore, intermodalism enhances flexibility, allowing businesses to choose the most suitable mode for each leg of the journey based on factors like cost, speed, and capacity requirements. Ultimately, it contributes to a more resilient and adaptable global supply chain.

The success of intermodalism hinges on well-coordinated infrastructure and standardized procedures. Efficient intermodal terminals serve as crucial hubs, facilitating smooth transitions between modes. These terminals must be strategically located, offering excellent access to major transportation routes and equipped with advanced handling equipment to minimize delays. Standardized containers and information systems play a vital role in reducing delays and enhancing security. Imagine a system where a single container smoothly moves from a train to a ship to a truck, all without ever being opened—this is the power of intermodal synergy. Effective regulatory frameworks are crucial to overcome cross-border challenges and streamline international shipments. This harmonization creates a level playing field for global businesses, fostering economic growth and creating opportunities for all stakeholders.

Global intermodal networks are intricate webs connecting various transportation hubs worldwide. Major ports, rail yards, and airports function as pivotal nodes within these networks, shaping global trade flows. Understanding these networks' structure is paramount for businesses aiming to leverage intermodal advantages. Analyzing key routes and potential bottlenecks provides insight into potential vulnerabilities and informs strategic decision-making. For instance, geographic constraints or infrastructure limitations in certain regions might impact efficiency. Likewise, economic factors like fuel prices and labor costs can influence the choice of transport mode and route optimization. Navigating these complexities requires a thorough grasp of global logistics and a sophisticated understanding of the interplay between transportation modes and economic forces.

Effective intermodal transportation depends heavily on advanced technologies. Real-time tracking systems provide crucial visibility into cargo location and status, allowing for proactive management and issue resolution. Predictive analytics, utilizing data on historical patterns and real-time conditions, enables optimized route planning, avoiding potential delays and minimizing transport costs. Blockchain technology offers increased security and transparency across the entire supply chain, while automation streamlines operations and reduces manual error, thereby boosting efficiency. The integration of these technologies significantly enhances operational efficiency and reduces the risk of delays or disruptions, creating a more reliable and cost-effective system. Companies that harness the power of these tools gain a competitive edge in the global marketplace.

The future of intermodal transport is intrinsically linked to sustainability. The industry faces increasing pressure to adopt environmentally friendly practices, and a transition towards greener transport modes and technologies is underway. The rise of electric vehicles, hybrid technologies and biofuels, alongside improvements in fuel efficiency and aerodynamic design are all contributing to a reduction in the sector's environmental footprint. Furthermore, innovations in materials science are contributing to lighter and more durable containers, while optimized logistics reduce unnecessary mileage, leading to significant reductions in greenhouse gas emissions. Companies and governments alike are increasingly prioritizing sustainable practices, incentivizing environmentally responsible operations and fostering a greener future for intermodal transport. This transition is not simply an environmental imperative, but a growing business necessity, as consumers and stakeholders increasingly demand sustainability within their supply chains. The future success of intermodal transport will be largely defined by its ability to balance economic viability with environmental responsibility.

1.1 Defining Intermodal Transport

Intermodal transport, at its core, is the seamless movement of goods utilizing multiple modes of transportation—rail, road, sea, and air—without handling the freight itself during transfers between modes. This integrated approach offers significant advantages over single-mode transportation, including cost savings, increased efficiency, and enhanced reliability. Think of it as a sophisticated relay race, where each leg is handled by a different runner, but the baton (the goods) is passed smoothly and quickly.

The key to successful intermodal transport lies in standardization. Uniform container sizes and handling procedures are crucial for efficient transfers between different transport modes. Without this, the system grinds to a halt, overwhelmed by incompatibility and delays. Imagine trying to fit square pegs into round holes—it simply won't work. Standardized procedures are essential for smooth transitions.

Effective intermodal transport necessitates a complex network of interconnected infrastructure. This includes well-equipped ports, strategically located rail yards, and efficient road networks. These nodes are interconnected, functioning as hubs within a larger system. The efficiency of this network directly impacts the speed and cost of transporting goods. Think of it as a circulatory system, where each component is vital for the overall health and function of the organism.

The economic benefits of intermodal transport are considerable. By optimizing the use of different transportation modes, businesses can reduce their overall logistics costs. This is particularly true for long-distance shipments where the combined efficiency of rail and sea transport surpasses that of road transport alone. The cost savings can be substantial, leading to a greater profit margin for businesses. The effects also reach consumers, offering more affordable products.

Furthermore, intermodal transport enhances supply chain resilience. By diversifying transportation options, businesses can mitigate risks associated with disruptions in a single mode of transport. A sudden strike by truck drivers, for example, would have a less devastating impact on a company that relies on an intermodal strategy. This flexibility adds robustness to supply chains, allowing them to withstand unforeseen circumstances.

However, challenges remain. One major hurdle is the lack of interoperability between different transport modes and systems. This often leads to delays and inefficiencies, highlighting the need for greater collaboration and standardization across the industry. A crucial step towards addressing these challenges is fostering better cooperation between various stakeholders, including carriers, terminal operators, and regulatory bodies.

Effective intermodal transport requires sophisticated logistical planning and management. Real-time tracking, optimized routing, and efficient scheduling are all essential for ensuring smooth and timely delivery. Advancements in technology, such as GPS tracking and predictive analytics, are playing an increasingly vital role in enhancing the efficiency and effectiveness of intermodal transportation systems. This sophisticated management is crucial for the success of the whole operation.

Governments play a pivotal role in fostering the growth of intermodal transport. Investment in infrastructure, the development of supportive regulatory frameworks, and the promotion of collaboration between different stakeholders are all critical for creating a conducive environment for the industry to flourish. This public-private partnership is a key factor for a sustainable and efficient system.

The future of intermodal transport is bright, driven by continued technological innovation and growing demand for efficient and sustainable logistics solutions. The integration of emerging technologies like autonomous vehicles and drone delivery has the potential to further revolutionize the industry, making it even more efficient and cost-effective. The potential of technological development is a key factor to consider for the future.

In conclusion, intermodal transport represents a powerful and versatile strategy for moving goods efficiently and cost-effectively. By understanding its core principles, addressing its challenges, and embracing technological advancements, stakeholders can unlock its full potential, ensuring the smooth flow of goods and the continued growth of global trade. The future of global logistics and the economy hinges on effective implementation.

1.2 Global Intermodal Networks

The global intermodal network isn't just a collection of disparate transportation systems; it's a complex, interconnected web, a living organism constantly adapting to the ever-changing demands of global trade. Think of it as a giant circulatory system, with arteries of railways, veins of highways, and capillaries of last-mile delivery networks, all working in concert to pump goods across continents. Each node, from bustling container ports to strategically located rail yards, plays a vital role in maintaining the flow. Disruptions in one area, such as a port strike or a rail bottleneck, can trigger ripple effects that impact the entire system.

Understanding this intricate network requires a grasp of its geography, its chokepoints, and its inherent vulnerabilities. Major port cities like Rotterdam, Shanghai, and Singapore serve as critical hubs, handling massive volumes of containerized goods. Similarly, strategically situated rail junctions and inland waterway terminals act as vital transfer points, facilitating the seamless transition of cargo between different modes of transport. But these very hubs, while essential for efficiency, also represent points of potential failure, making them targets for both logistical challenges and security threats. Robust contingency planning and collaborative partnerships between stakeholders are therefore paramount.

Efficient coordination is the lifeblood of any successful intermodal network. This involves standardized procedures, clear communication channels, and the consistent application of tracking and tracing technologies. Imagine a shipment traversing the globe, moving from a factory in China via sea, then rail to an inland distribution center, and finally by truck to its ultimate destination in Europe. Each leg of the journey requires meticulous planning, precise timing, and efficient handling. Any delay or mismatch in procedures can lead to significant disruptions, causing delays, increased costs, and potential damage to goods.

The role of technology in enhancing the efficiency and resilience of these networks cannot be overstated. Real-time tracking systems, sophisticated scheduling algorithms, and predictive analytics are increasingly essential tools for optimizing logistics and minimizing downtime. These technologies not only improve efficiency but also enhance transparency and accountability throughout the supply chain, benefiting all stakeholders, from manufacturers and shippers to consumers. The strategic use of data analytics is helping to identify bottlenecks, predict disruptions, and proactively manage risks.

Furthermore, the environmental impact of intermodal transport is a significant concern demanding careful attention. The industry needs to embrace sustainable practices to minimize its carbon footprint. This means promoting the use of fuel-efficient vehicles, investing in renewable energy sources, and developing environmentally friendly packaging solutions. Regulations and incentives can play a crucial role in encouraging the adoption of greener technologies and practices. Companies are starting to see the value in sustainable practices, not just as a matter of corporate social responsibility, but also as a competitive advantage. Consumers, too, are showing a growing preference for environmentally conscious companies and products.

Looking ahead, the global intermodal network faces both opportunities and challenges. E-commerce growth, for instance, is driving demand for faster and more efficient delivery services. Simultaneously, the increasing need for supply chain resilience in the face of geopolitical instability and natural disasters necessitates the development of robust and adaptable systems.

These issues require a collaborative response, with governments, industry stakeholders, and technological innovators working together to shape the future of global freight movement. The future of efficient and effective global trade depends on a concerted global effort. It's a challenge we must meet to ensure the continued flow of goods across continents. The consequences of failure are simply too great to ignore. Innovative solutions and collaborative approaches will be crucial in navigating this dynamic landscape. Investing in infrastructure upgrades, promoting sustainable practices, and harnessing the power of technology will be vital. The interconnectedness of the modern world highlights the importance of a seamless global intermodal network. Success in the years ahead depends on a global commitment to efficiency, sustainability, and resilience.

2 Railroad Integration Strategies

Efficient rail yard operations are paramount for seamless intermodal integration. Think of a well-oiled machine; every component must function smoothly. This necessitates optimized track layouts minimizing unnecessary switching maneuvers, advanced signaling systems preventing delays, and sophisticated scheduling software to manage train arrivals and departures with precision. Investing in automated sorting systems, capable of rapidly processing high volumes of containers, dramatically reduces handling time and labor costs. Furthermore, the integration of real-time data analytics allows for proactive identification of bottlenecks and potential disruptions, enabling swift corrective actions. Effective communication networks connecting yard personnel, dispatchers, and other stakeholders are equally vital for maintaining efficient operations. Ultimately, these strategies ensure a continuous flow of goods, minimizing delays and maximizing throughput.

High-speed rail significantly enhances the competitiveness of intermodal transport. Faster transit times directly translate to reduced delivery times, a key advantage in today's time-sensitive global marketplace. This speed advantage allows for the creation of more efficient and responsive supply chains. For instance, businesses can rely on shorter lead times, potentially reducing inventory costs and enhancing customer satisfaction. However, the development of high-speed rail networks often necessitates substantial investment in infrastructure. Careful consideration must be given to the economic feasibility, taking into account construction costs, operating expenses, and projected revenue streams. Detailed environmental impact assessments are also crucial to mitigate potential negative consequences. Moreover, effective integration with existing rail networks and other transportation modes is critical to maximizing the benefits of this faster transit time, ensuring seamless connections.

Effective intermodal terminal design is critical for success. Strategic location is paramount, considering proximity to major transportation hubs, accessibility for various modes of transport, and sufficient space for operations and future expansion. A well-designed terminal promotes efficient transfers between rail, road, and potentially maritime or air transport. This implies well-planned layouts that optimize container flow, minimizing congestion and unnecessary handling. Adequate infrastructure, such as sufficient loading docks, cranes, and storage areas, is essential to accommodate peak demand and ensure smooth operations. The incorporation of technological advancements, such as automated guided vehicles (AGVs) and real-time tracking systems, further enhances efficiency and minimizes human error. Furthermore, security features should be integrated into the design to protect goods and ensure the safety of personnel. Ultimately, a well-designed terminal acts as a vital node in the intermodal network.

The strategic development and implementation of these railroad integration strategies directly impact the overall efficiency and competitiveness of intermodal transport systems. By focusing on improvements within rail yards, embracing high-speed rail, and designing efficient intermodal terminals, we can create a more seamless, faster, and cost-effective transportation network. This network in turn fosters economic growth, facilitates international trade, and improves the overall quality of goods and services available to consumers globally. This is not only about the engineering aspects but also about how we weave together different elements to streamline operations and provide more efficient service. The future of logistics is inherently tied to these improvements; the faster and more efficient the transport, the better the service. This means investing in the right infrastructure and technology that allows for a flow of goods, not a bottleneck. Therefore, the importance of this integration cannot be overstated.

2.1 Rail Yard Efficiency

Efficient rail yard operations are crucial for seamless intermodal transport. Delays in a single yard can ripple through the entire system, causing significant economic losses and frustrating customers. Therefore, optimizing yard processes demands a multifaceted approach.

Improving rail yard efficiency begins with effective planning and design. This includes strategically locating yards to minimize transportation distances and incorporating modern technologies to streamline operations. Consideration should be given to the yard's layout to optimize the flow of trains and minimize congestion. This might involve utilizing advanced track designs, optimizing the placement of classification tracks, and ensuring sufficient space for storage and maintenance. Proper planning reduces the need for costly redesigns and operational disruptions down the line. A well-designed yard should accommodate anticipated growth while maintaining flexibility to adapt to changing transportation demands. The choice of technologies to incorporate should be made with care, considering both long-term cost and integration with existing systems.

Next, efficient train scheduling and management are paramount. Utilizing advanced software systems helps predict and manage potential delays. Real-time tracking of train locations and predicted arrival times allows for proactive adjustments to the schedule, minimizing disruptions. Implementing just-in-time scheduling helps ensure that trains arrive precisely when needed, reducing unnecessary waiting times and improving overall yard efficiency. This precision scheduling minimizes congestion and improves the utilization of track space and personnel. Sophisticated algorithms can be used to analyze historical data and predict future demands, leading to more effective resource allocation.

Technology plays a vital role in modernizing rail yards. Automated switching systems can significantly reduce the need for manual labor, improving both speed and safety. These systems can precisely control the movement of individual cars, accelerating the process of assembling and disassembling trains. Moreover, sensors and data analytics can provide real-time information on track conditions, train movements, and equipment status. This allows for proactive maintenance and quick responses to unexpected events, preventing delays. Investments in real-time data visualization tools and enhanced communication systems can further improve coordination among yard personnel and other stakeholders in the intermodal network, leading to more informed decision-making and a smoother flow of goods.

Human capital is just as important as technology. Well-trained personnel are essential for safe and efficient rail yard operations. This requires investing in comprehensive training programs that focus on both technical skills and safety procedures. Empowering workers through clear communication channels and participatory management styles fosters a sense of ownership and responsibility, leading to increased efficiency. Effective teamwork and clear communication between personnel operating different components of the yard are crucial in the smooth execution of processes, minimizing delays caused by miscommunication or coordination issues. Investing in employee well-being, including promoting a safe and supportive work environment, results in higher morale and thus improved productivity.

Finally, collaboration with other stakeholders in the intermodal transportation system is crucial for maximizing rail yard efficiency. This involves working closely with trucking companies, shipping lines, and other partners to ensure seamless transitions between modes. Establishing clear communication channels and standardized procedures for exchanging information and coordinating schedules is paramount. Efficient data sharing facilitates optimized scheduling and reduces delays caused by mismatches between modes of transport. Regular meetings and collaborative planning sessions can help proactively identify and resolve potential bottlenecks before they impact operations. Open and consistent communication creates a culture of trust and cooperation, leading to smoother integration within the broader intermodal network. A collaborative approach ensures that the efficiency of the rail yard contributes to the overall efficiency of the entire intermodal transportation chain.

2.2 High-Speed Rail Links

High-speed rail presents a compelling opportunity to significantly enhance intermodal transport networks. Its speed advantage drastically reduces transit times, making it a highly attractive option for time-sensitive goods and passengers. This speed translates directly into economic benefits, reducing inventory holding costs for businesses and offering consumers faster travel options. Integrating high-speed rail necessitates careful consideration of the surrounding infrastructure. Effective interchanges with other transport modes—road, sea, and air—are crucial. These interchange points must be designed with seamless transfers in mind, minimizing delays and handling complexities. Consider the example of a well-designed high-speed rail station connected to a major airport via a dedicated, high-capacity bus service. Passengers can easily transfer between modes, ensuring a smooth and efficient journey.

The construction and operation of high-speed rail lines demand substantial investment. However, the long-term economic returns can be significant, stimulating economic growth in surrounding regions and boosting overall national productivity. Job creation throughout the project lifecycle—from construction to operation and maintenance—is substantial. Moreover, efficient freight transport via high-speed rail can facilitate just-in-time manufacturing strategies, leading to leaner operations and reduced waste. To achieve these economic benefits, thorough feasibility studies are essential. These studies must accurately assess potential ridership and freight volumes, alongside the projected costs and financial returns. A comprehensive analysis of environmental impacts, including carbon emissions and land use, is equally crucial. The potential for reduced road congestion and associated air pollution benefits from the shift in modal share should also be explored. Beyond economic viability, the success of high-speed rail integration hinges on effective governance and policy. Clear regulatory frameworks are needed to govern the construction, operation, and safety of high-speed rail lines. These frameworks must encompass issues such as land acquisition, environmental protection, and safety standards. Furthermore, strong public-private partnerships can facilitate the necessary financial investment and expertise. Government subsidies and incentives can play a crucial role in making these projects economically feasible, especially in the initial stages. International collaboration is also essential to promote standardization and interoperability across national borders. This is particularly important for routes that traverse multiple countries, ensuring seamless cross-border travel and freight transport. Effective communication and public engagement are essential to address potential concerns and secure public support for these projects. Transparency in planning and execution can help mitigate potential opposition from local communities.

Technological advancements play a vital role in optimizing high-speed rail operations. Advanced signalling systems, such as ETCS (European Train Control System), enhance safety and increase capacity. Real-time passenger information systems improve the travel experience, while sophisticated maintenance management systems help to reduce downtime and ensure operational efficiency. The integration of digital technologies allows for better data analysis, improving route planning and optimizing resource allocation. Furthermore, exploring possibilities for autonomous or semi-autonomous train operation can further enhance efficiency and safety in the future. These technological improvements are not just about speed and efficiency; they can significantly improve the overall passenger and freight experience. Data-driven decision-making, facilitated by these advancements, allows for continuous improvement and adaptation to changing needs.

In conclusion, integrating high-speed rail into intermodal networks offers tremendous potential for economic growth, improved efficiency, and enhanced connectivity. However, realizing this potential requires careful planning, substantial investment, effective governance, and strategic deployment of advanced technologies. A holistic approach, considering economic, environmental, and social factors, is essential for successful integration. Successfully navigating the complexities of implementation will create a significantly more efficient and resilient global transportation system for both passengers and goods. The long-term benefits far outweigh the initial challenges, positioning high-speed rail as a cornerstone of future intermodal transportation strategies. By strategically integrating high-speed rail with other transport modes, we can build a truly interconnected and sustainable global transportation network. The potential for economic benefits and environmental improvements is significant, making the investment a compelling proposition for governments and private investors alike. Careful planning and execution are paramount to ensuring its success.

2.3 Intermodal Terminal Design

Efficient intermodal terminal design hinges on a deep understanding of throughput requirements. Forecasting future cargo volumes, considering both current trends and projected growth, is crucial. This informs the scale of the terminal, determining the number of tracks, loading docks, and storage areas needed. Ignoring this initial step often leads to capacity constraints and operational bottlenecks down the line, significantly impacting efficiency and profitability. A robust design should anticipate future expansion, allowing for modular additions without major disruptions to ongoing operations. Careful consideration of the land's geographical features and environmental factors also play a vital role, including the assessment of soil stability for foundations and the impact on the surrounding environment. Location selection is paramount. Proximity to major transportation networks—rail lines, highways, waterways, and airports—is essential for seamless modal transfer. Minimizing distances between modes reduces transit time and lowers costs, thereby enhancing the terminal's overall competitiveness. Furthermore, access to efficient infrastructure, including utilities, communication networks, and skilled labor, should be a major deciding factor. These are crucial elements for maintaining smooth operations and attracting potential tenants and customers. Ignoring these logistical aspects can lead to significant delays and operational headaches later. Analyzing transportation costs, local regulations, and available workforce skills can help identify the optimal site.

Layout planning necessitates optimizing workflow. The terminal's layout should facilitate a smooth flow of goods, minimizing congestion and unnecessary movements. Strategic placement of loading docks, storage areas, and inspection points minimizes handling time and improves efficiency. Furthermore, it is crucial to anticipate the types of cargo handled, adapting the design accordingly. For example, a terminal handling heavy machinery requires different infrastructure compared to one specializing in containerized goods. Therefore, careful consideration of equipment needs – from cranes and forklifts to specialized handling equipment – is vital to design a layout that supports efficient cargo movement. This is a critical aspect often overlooked in initial design phases, but one that fundamentally impacts the terminal's operational effectiveness.

Technological integration is not merely an add-on but a fundamental element of modern terminal design. Real-time tracking systems, automated gate systems, and yard management systems (YMS) are not luxuries but necessities. These technologies improve efficiency, enhance security, and offer better visibility throughout the logistics chain. They streamline operations, reduce manual handling, and improve overall throughput. Furthermore, data analytics, when integrated with these technologies, can provide insights into operational bottlenecks, allowing proactive measures to improve efficiency. The investment in robust IT infrastructure, including high-speed internet access and reliable communication networks, is therefore critical for effective technological integration and optimal operational performance. This is no longer merely an investment in technology but an investment in the future of competitive edge.

Sustainability considerations are no longer optional but essential. Environmental impact assessments are crucial, exploring energy-efficient technologies, waste management strategies, and minimizing carbon emissions. These factors are increasingly important for attracting environmentally conscious clients and complying with increasingly stringent environmental regulations. The choice of building materials, lighting systems, and transportation modes within the terminal itself all contribute to its environmental footprint. Therefore, environmentally friendly practices should be integrated into all aspects of the terminal's design and operation, not just as an afterthought but as core operational priorities. These measures contribute to a greener terminal and reflect a commitment to sustainable practices.

Effective security measures are paramount. A well-designed terminal includes robust security systems to protect against theft, vandalism, and terrorism. This involves integrating technologies such as CCTV surveillance, access control systems, and perimeter security measures. Moreover, collaboration with law enforcement agencies and implementation of strict security protocols are necessary to ensure the safe and secure operation of the facility. These systems need to be seamlessly integrated into the overall terminal operations, not as an afterthought, but as an integral part of the daily functioning. This integrated security approach ensures not just the protection of cargo, but also the safety of personnel and the smooth flow of operations.

Finally, adaptable design allows for future adjustments and expansions to meet evolving needs. The terminal should not be viewed as a static entity, but rather a dynamic system capable of adapting to changes in cargo volume, types of cargo, and technological advances. Flexible layouts, scalable infrastructure, and modular designs are key to ensuring the long-term viability and competitiveness of the intermodal terminal. This allows for accommodating future growth and technological updates without the need for major and costly reconstructions. Investing in an adaptable design is an investment in the future, ensuring the longevity of this important link in global supply chains.

3 Maritime Port Operations

Efficient container handling is paramount in modern port operations. Sophisticated technologies like automated stacking cranes and automated guided vehicles (AGVs) are increasingly employed to accelerate the movement of containers, minimizing congestion and optimizing throughput. Real-time tracking systems provide precise location data, facilitating faster retrieval and reducing the risk of misplacement. This leads to significant cost savings and improved customer satisfaction, as goods reach their destinations more quickly and reliably. The integration of these systems requires careful planning and coordination, along with substantial investment in infrastructure and skilled labor. Furthermore, the effective use of these technologies necessitates robust data management and sophisticated software for coordinating the intricate choreography of container movement within the port. Success hinges on seamless communication between different systems and departments, allowing for a coordinated and synchronized flow of goods.

Robust port infrastructure is the backbone of efficient maritime operations. Deep-water berths allow the docking of larger vessels, increasing cargo capacity. Well-maintained roads, rail connections, and sufficient warehousing space are crucial for efficient cargo transfer between the port and other transportation modes. The design and construction of such infrastructure must consider factors like seismic activity, potential flooding, and the environmental impact. Investing in modern infrastructure not only enhances efficiency but also strengthens a port's competitiveness in the global market. Neglecting this area can lead to bottlenecks, delays, and increased costs. A well-planned and integrated infrastructure strategy considers future expansion needs and incorporates sustainable design principles, minimizing the environmental footprint while optimizing operational efficiency. Furthermore, adequate security measures within the infrastructure are essential to protect against theft and sabotage. Strategic planning and investment in port infrastructure are vital for sustained economic growth and international trade.

Smart port technologies are revolutionizing maritime operations, enhancing efficiency, security, and sustainability. These technologies include advanced sensors and IoT devices for monitoring equipment, optimizing energy consumption, and detecting potential hazards. Artificial intelligence (AI) and machine learning (ML) algorithms are used for predictive maintenance, streamlining port processes, and improving resource allocation. Data analytics provide valuable insights into operational inefficiencies, helping identify areas for improvement and optimization. By integrating these technologies, ports can significantly reduce operational costs, enhance security protocols, and minimize their environmental footprint. The adoption of smart port technologies is an ongoing process, requiring collaboration between port authorities, technology providers, and logistics operators. The ultimate goal is a fully integrated, intelligent ecosystem that optimizes every aspect of port operations, making them faster, safer, and more sustainable. Maritime security is a critical concern, requiring multifaceted strategies. This includes robust physical security measures like access control systems, CCTV surveillance, and perimeter fencing. Cybersecurity protocols are equally crucial, protecting port operations from cyber threats and data breaches. Effective collaboration between port authorities, law enforcement, and intelligence agencies is vital for coordinating security efforts and sharing information. International cooperation is also essential for tackling transnational crime and terrorism. Regular security assessments and drills are necessary for identifying vulnerabilities and improving response capabilities. Investing in advanced security technologies and training programs is crucial for maintaining a secure and efficient port environment. A proactive approach to maritime security safeguards trade, protects infrastructure, and ensures the safety of personnel and cargo. It is not just about physical security but also about safeguarding sensitive data and critical infrastructure.

Logistics optimization is key to efficient port operations, focusing on streamlining processes and minimizing delays. This involves optimizing vessel scheduling, berthing assignments, and container handling procedures. Efficient yard management and effective coordination with trucking companies are crucial for optimizing the movement of cargo both within the port and beyond. The use of advanced planning and scheduling software can significantly improve efficiency and reduce costs. Effective communication and collaboration between all stakeholders are crucial for optimizing logistics. Data analytics can help identify and address bottlenecks, leading to improved overall efficiency. Regular review and adjustment of logistics plans, based on performance data and real-time conditions, are essential for adapting to changing demands and ensuring optimal performance. The ultimate aim is to create a seamless flow of goods, minimizing dwell time and maximizing throughput.

3.1 Container Handling

Efficient container handling is paramount to the smooth operation of any intermodal network. It's not just about moving boxes; it's about optimizing the entire logistical chain, from the moment a container leaves its origin to its final destination. Failure at this stage ripples through the system, causing delays, increasing costs, and frustrating customers. Consider the sheer volume of goods transported globally—a smoothly functioning container handling system is essential for global commerce.

The choice of handling equipment significantly impacts efficiency. Reaching for the right equipment—from gantry cranes to reach stackers, depending on the scale and type of operation—is critical. Larger ports often utilize automated systems, showcasing advancements in robotics and control systems, significantly improving speed and precision while minimizing human error. Smaller ports, however, might rely on more conventional methods, demanding a keen understanding of manual processes for optimal efficiency. Careful planning and investment in appropriate technology, tailored to the specific needs of each port, are vital for success.

Effective container stacking is more than just piling boxes. Careful planning is needed to optimize space utilization, ensuring easy retrieval, and minimizing the risk of damage during stacking and retrieval. Strategic placement considers the weight and fragility of the goods, as well as the size and type of container. Efficient stacking methodologies and training for personnel are important elements to mitigate damages and improve productivity. Proper labeling and a well-organized inventory system allow for swift identification and retrieval, preventing delays and reducing wasted time. The focus must always be on safety and efficiency.

Beyond the equipment, personnel training plays a crucial role. Well-trained personnel are the backbone of a successful operation, capable of operating equipment safely and efficiently. Their skills contribute significantly to faster processing and reduced errors. Regular training and certification programs that maintain skills and update personnel on safety regulations and the latest technologies are essential. Regular safety drills and refresher courses instill safe work practices, ensuring a secure environment for workers and the smooth, accident-free running of the port. The continuous improvement and well-being of employees directly influences the efficiency and reliability of the system.

The integration of technology is transforming container handling. Real-time tracking systems provide transparency, allowing for efficient monitoring of containers throughout their journey. Predictive analytics can optimize processes by anticipating potential bottlenecks and allowing for proactive adjustments. Automated guided vehicles (AGVs) and other automated systems are improving efficiency and minimizing the need for human intervention in repetitive tasks. Moreover, intelligent scheduling systems utilize data-driven insights for optimizing the movement of containers, reducing congestion and improving overall throughput. Embracing technological advances is essential for remaining competitive in a fast-paced industry.

Security is a vital aspect of container handling. Robust security protocols, incorporating technological advancements like CCTV surveillance and advanced identification systems, are crucial in preventing theft and other security breaches. Thorough inspections and screening processes are important to ensure that only authorized personnel and cargo enter the facility, safeguarding against contraband and other threats. Port authorities often collaborate with law enforcement agencies to share intelligence and coordinate security measures. Investing in strong security measures is not only vital for business but also for maintaining the integrity and efficiency of the entire supply chain.

The optimization of container flow involves several strategies. Efficient yard management systems, using real-time data to track containers, and optimize their placement for streamlined movement, are crucial. Careful planning and coordination are required to minimize unnecessary movement and optimize the flow of containers through the port. Effective communication and collaboration among stakeholders, such as shipping lines, trucking companies, and customs officials, are vital for ensuring the smooth and timely movement of goods. Improving overall throughput requires effective collaboration and management of the process as a whole.

3.2 Port Infrastructure

Efficient port infrastructure is the backbone of global maritime trade, acting as crucial hubs where goods transition between sea and land transport. A well-designed port significantly reduces transit times, minimizes congestion, and lowers overall transportation costs. This translates directly into economic benefits for businesses and nations alike. Consider, for example, the impact of a poorly designed port – extended waiting times for ships, increased fuel consumption, and potential delays in the supply chain, all leading to significant financial losses. Conversely, an efficient port attracts more shipping lines, boosts regional economies, and contributes to a more resilient global supply chain. Investing in robust port infrastructure is, therefore, a strategic move with far-reaching positive consequences.

The design and construction of modern ports demand careful consideration of numerous factors. Depth and width of waterways are critical, ensuring sufficient capacity for large vessels. Sophisticated navigational systems, including GPS-aided vessel tracking and automated berth allocation, are essential for optimizing vessel traffic flow and preventing collisions. Efficient dredging operations are necessary to maintain channel depths, especially in areas susceptible to sedimentation. The integration of advanced technologies, such as real-time monitoring of water levels and weather conditions, can further enhance safety and operational efficiency. These are just some of the elements that combine to define a technologically advanced port.

Beyond the waterways themselves, the landside infrastructure plays a vital role. Efficient container handling facilities, including strategically placed cranes and automated stacking systems, are crucial for expediting the movement of goods. Well-planned road and rail connections, minimizing the distance between the port and inland distribution centers, are also crucial. This seamless integration of different modes of transport is essential for reducing overall transportation times and costs. A lack of efficient landside infrastructure creates significant bottlenecks, leading to delays, congestion, and increased operational expenses for businesses. Investing in well-developed landside infrastructure is as important, if not more so, than improvements to the seaward elements of the port.

Moreover, the environmental impact of port operations cannot be overlooked. Ports are often located in ecologically sensitive areas, and their operations can have a significant impact on local ecosystems. Measures to mitigate environmental damage are therefore essential. This can include the implementation of noise and air pollution control measures, the careful management of wastewater discharge, and the adoption of sustainable practices throughout port operations. Furthermore, considering the long-term effects on the surrounding environment should be a central part of the design and development process. Environmental considerations are no longer a matter of optional add-ons, but rather a fundamental part of responsible port development.

Finally, effective port management is equally crucial. This involves implementing streamlined procedures for customs clearance, coordinating the movement of goods through the port, and ensuring the safety and security of port facilities and personnel. Effective management systems can significantly improve the efficiency and productivity of the port. Investing in advanced information technology systems, integrating data from different sources, and using predictive analytics can further enhance port management capabilities. The optimal management system is one that leverages technological advancements to optimize port operations and respond efficiently to changing market demands. This holistic approach, encompassing infrastructure, technology, and management, is vital for creating a truly efficient and sustainable port. It is the integration of these different aspects which allows the port to fulfil its role as a key element in the global intermodal network.

3.3 Smart Port Technologies

Smart ports leverage technology to optimize every aspect of operations, from vessel scheduling to cargo handling. This isn't just about efficiency; it's about creating a more resilient, secure, and environmentally responsible maritime industry. Consider, for instance, the implementation of advanced container tracking systems using RFID tags and IoT sensors. These technologies provide real-time visibility of cargo location and condition, improving inventory management and reducing the risk of theft or damage. Furthermore, predictive analytics, powered by machine learning, can forecast container arrival times with greater accuracy, allowing for smoother port operations and reducing congestion.

Digital twin technology offers a virtual representation of the entire port infrastructure. This allows for simulating different scenarios, such as changes in traffic flow or equipment malfunction, which helps identify potential bottlenecks and optimize resource allocation before they impact actual operations. Such foresight is invaluable. Moreover, the port's ability to adapt to unforeseen circumstances, like severe weather events or unexpected surges in cargo volume, is significantly enhanced. This translates directly to improved operational resilience and reduced disruptions. By running virtual simulations, port authorities can fine-tune their response plans and ensure a smoother transition during unexpected events.

Automated guided vehicles (AGVs) and other automated systems are revolutionizing cargo handling within port terminals. These systems can operate continuously, 24/7, handling containers with greater speed and precision than manual methods. Think of the significant reduction in labor costs and increased throughput they deliver. More importantly, however, automation diminishes the risk of human error, leading to fewer accidents and improved safety for port workers. The integration of robotics and AI-powered systems also opens up opportunities for optimizing cargo stacking, improving space utilization within the terminal, and reducing the environmental footprint through optimized fuel consumption of AGVs.

Security is paramount in modern ports. Smart port technologies enhance security measures through advanced surveillance systems, including CCTV cameras with intelligent video analytics and biometric access control. These systems not only deter criminal activity but also enable rapid response to any security breaches. The use of blockchain technology provides an added layer of security for tracking and managing cargo documentation, ensuring transparency and reducing the risk of fraud or counterfeiting. This robust digital approach reduces bureaucratic hurdles and creates trust among all stakeholders.

Data analytics plays a critical role in optimizing port operations. By collecting and analyzing data from various sources, such as vessel arrival times, cargo volume, weather conditions, and equipment performance, port authorities can gain valuable insights into their operations. This data-driven approach enables them to identify areas for improvement, optimize resource allocation, and make more informed decisions regarding infrastructure development and capacity planning. Data analytics empowers them to proactively address challenges, minimizing operational costs and maximizing efficiency. For example, by analyzing historical data, ports can predict peak demand periods and adjust their resources accordingly to prevent bottlenecks and maintain optimal operational flow.

Integration with other transportation modes is crucial for smart port efficiency. Smart ports are not isolated entities; rather, they act as crucial hubs within broader intermodal transportation networks. Seamless data exchange between ports and other modes, such as rail and road, is essential for facilitating efficient cargo movement throughout the supply chain. Real-time information sharing allows for optimized scheduling, reduces waiting times, and minimizes delays, leading to significant cost savings for shippers and freight forwarders. Effective collaboration between different transportation modes necessitates a unified communication platform, where data flows smoothly among all participants.

Sustainable practices are an integral aspect of smart port development. Smart ports aim to minimize their environmental impact through the implementation of green technologies, such as renewable energy sources, energy-efficient equipment, and waste management systems. Reducing emissions and improving energy efficiency contribute to a cleaner, more sustainable maritime industry. Furthermore, smart ports contribute to a reduction in overall transportation costs because of fuel savings, thus lowering overall environmental impacts linked to the transportation process. The shift towards sustainable practices not only enhances a port's environmental performance but also enhances its appeal to environmentally conscious businesses.

The future of smart ports lies in the continued development and integration of advanced technologies, fostering greater collaboration and data sharing throughout the supply chain. This evolution will require substantial investments in infrastructure, technology, and workforce training. Nevertheless, the long-term benefits of adopting smart port technologies, in terms of increased efficiency, enhanced security, improved sustainability, and economic growth, are immense. As the global demand for maritime transportation continues to rise, smart port technologies are not merely a luxury; they are a necessity.

3.4 Maritime Security

Maritime security is paramount for the smooth functioning of global intermodal transport. A single security breach at a major port can create ripple effects, causing significant delays, economic losses, and even geopolitical instability. Robust security measures are therefore not merely desirable, but absolutely essential for the continued efficiency and viability of the entire intermodal system. Effective security protocols require a multi-layered approach, encompassing physical security, technological advancements, and strong international collaboration.

Implementing comprehensive physical security measures is the foundation of maritime security. This involves upgrading port infrastructure with advanced surveillance systems, including high-definition CCTV cameras, motion detectors, and perimeter fencing reinforced with intrusion detection systems. Regular security patrols by trained personnel, coupled with access control systems that verify the identity of every individual and vehicle entering the port, are crucial. Investing in robust lighting systems to eliminate dark areas where criminal activity could take place is also paramount. Furthermore, the use of specialized detection equipment like x-ray scanners and explosive detection devices is crucial for the early identification of potential threats. These systems need to be integrated and monitored by a central control room, allowing for real-time monitoring and immediate response to any security incident. The efficiency and effectiveness of these systems ultimately rely on well-trained personnel and efficient protocols for quick intervention.

Beyond physical infrastructure, technological advancements are revolutionizing maritime security. Real-time data tracking and analysis systems, integrating data from various sources such as ships' Automatic Identification Systems (AIS), CCTV feeds, and other sensor networks, provide a comprehensive overview of port activities. This enables proactive identification of suspicious behavior and allows for timely intervention. Cybersecurity is equally crucial, as ports are increasingly reliant on computer systems to manage operations. Protecting these systems from cyberattacks is essential to prevent disruption and data breaches. Blockchain technology, with its immutable ledger system, presents opportunities for enhanced transparency and traceability of goods, improving the ability to track and prevent the movement of contraband. The implementation of these technologies, however, requires significant investment and expertise in both hardware and software infrastructure, along with dedicated staff to manage and maintain the systems. The integration of these advanced systems needs to be carefully planned and executed to ensure seamless operation and avoid creating new vulnerabilities.

International cooperation is crucial for effective maritime security. Sharing information across nations is essential to detect and prevent transnational criminal activities such as smuggling, piracy, and terrorism. Collaborative efforts in intelligence gathering, information sharing, and joint security operations can significantly enhance the effectiveness of individual national efforts. Harmonizing security standards and protocols across different ports and countries is important for seamless integration of security procedures within the global intermodal network. This collaboration must include not only governmental agencies but also private sector stakeholders, including shipping lines, port operators, and logistics providers. Effective communication and information-sharing platforms are crucial for creating a comprehensive, collaborative network that enhances maritime security worldwide. Joint training exercises and capacity-building initiatives among different nations are also essential for enhancing expertise and coordination in responding to security challenges.

Effective maritime security necessitates a combination of strong physical security measures, cutting-edge technology, and effective international cooperation. By strategically investing in these areas, the global intermodal transport system can significantly mitigate risks and ensure the safe and efficient flow of goods across borders. While challenges such as evolving criminal tactics and technological advancements continually test the effectiveness of existing measures, continuous adaptation, innovative approaches and global collaborations are essential for maintaining a secure and resilient maritime transportation ecosystem. Failure to adopt a comprehensive and proactive approach puts the entire intermodal system at significant risk, impacting global trade and supply chains. Therefore, constant vigilance, investment in innovative solutions, and robust international cooperation are not simply worthwhile goals; they are essential for the survival and success of the entire intermodal transport network. The future of intermodal transport is inextricably linked to the continued strength and effectiveness of its maritime security posture.

3.5 Logistics Optimization

Efficient port operations are crucial for the success of any intermodal system. Consider, for instance, the sheer volume of containers that pass through a major port daily; optimizing their movement is not simply about speed, but about minimizing congestion, reducing delays, and lowering costs across the entire supply chain. This requires a multifaceted approach, encompassing technological innovation and strategic planning. Investing in automated container handling systems, such as automated guided vehicles (AGVs) and remote-controlled cranes, significantly improves throughput and reduces the reliance on manual labor, leading to greater efficiency and improved safety. The implementation of sophisticated software systems for real-time tracking and scheduling is equally important. These systems enable efficient allocation of resources, predicting potential bottlenecks, and allowing for proactive adjustments. Such advanced tools enhance decision-making and minimize delays caused by unforeseen events.

Beyond technology, effective port infrastructure plays a vital role. This includes well-designed layouts that streamline container flow, ensuring sufficient space for storage and handling, and investing in robust infrastructure like adequate berthing facilities, efficient rail and road connections, and strategically placed warehouses. These elements create a smooth and seamless transfer of goods between different transportation modes, ensuring the quick and cost-effective movement of goods. Effective port management also necessitates strong collaboration between various stakeholders – port authorities, shipping lines, trucking companies, and customs officials – all of whom must work together to minimize bureaucratic hurdles and facilitate swift cargo processing. A well-coordinated system where information flows freely and responsibilities are clearly defined is indispensable for efficient operations. The strategic positioning of ports, taking into account geographical factors, trade routes, and market demands, is another critical aspect of optimization.

Logistics optimization in the maritime sector extends beyond the physical aspects of port operation. It also encompasses the strategic planning and management of cargo flows. This involves careful consideration of factors like shipment consolidation, route optimization, and inventory management. Consolidating smaller shipments into larger ones, for example, reduces transport costs by maximizing container utilization. Utilizing advanced route planning software that considers factors such as weather conditions, traffic congestion, and canal transit times allows for the selection of the most efficient routes, ultimately reducing travel time and fuel consumption. Effective inventory management, including accurate forecasting of demand and precise inventory tracking, minimizes storage costs and prevents stockouts or overstocking, both of which can significantly disrupt operations and impact profitability. The combination of these carefully considered actions dramatically improves the overall efficiency and reduces the environmental impact associated with maritime shipping.

Furthermore, incorporating predictive analytics is an increasingly crucial aspect of maritime logistics optimization. By analyzing historical data, current conditions, and predictive models, businesses can forecast future demand, optimize vessel scheduling, and proactively address potential disruptions. Such data-driven insights can help to anticipate and mitigate delays caused by factors like inclement weather or unexpected congestion, thereby improving on-time delivery rates and reducing operational costs. Implementing a robust reporting system that tracks key performance indicators (KPIs), including vessel utilization rates, cargo handling times, and delivery performance, allows for continuous monitoring and evaluation of efficiency and serves as a valuable basis for ongoing improvement. This data-driven approach not only identifies areas for improvement but also aids in implementing solutions, adapting to changing market dynamics and improving overall operational excellence. The integration of these various elements within a comprehensive logistics strategy represents a crucial step toward maximizing efficiency and minimizing disruptions within maritime port operations.

Finally, consider the implementation of smart port technologies. These advancements, ranging from sensor networks and automated systems to advanced data analytics platforms, offer significant opportunities to optimize port operations and enhance efficiency. Real-time data gathered from sensors embedded in containers, ships, and infrastructure provides valuable insights into cargo location, movement, and environmental conditions. This granular level of data allows for precise tracking of goods, proactive identification of potential problems, and efficient management of resources. The integration of Internet of Things (IoT) devices enables remote monitoring of equipment, allowing for predictive maintenance, reducing downtime, and optimizing the lifecycle of assets. Such predictive maintenance minimizes unplanned downtime, reducing operational costs and ensuring the smooth operation of vital port equipment. The intelligent utilization of these technologies is paramount in optimizing maritime port logistics and increasing overall efficiency.

4 Road Freight Collaboration

Effective road freight collaboration is crucial for a seamless intermodal system. Think of it as the final, often most intricate, piece of a complex puzzle. Getting this right is essential for on-time deliveries and satisfied customers; failure here can unravel the entire operation. The key lies in optimizing the interface between trucks and other modes, primarily rail, minimizing delays and cost overruns.

Efficient truck-rail transfers demand careful planning and coordination. Consider the design of intermodal terminals: easy access for trucks, swift and safe unloading and loading processes, and streamlined paperwork. This isn't just about physical infrastructure; effective communication systems are just as important – real-time tracking, efficient scheduling software, and clear communication protocols between trucking companies and rail operators are non-negotiable. Think about the potential for delays caused by inadequate communication: a single missed notification can cascade into a chain of problems, impacting delivery schedules and potentially causing significant financial losses.

Last-mile delivery presents unique challenges. Unlike long-haul transport, this stage involves navigating congested urban environments and dealing with a myriad of individual delivery points. Solutions include advanced routing algorithms which consider real-time traffic, efficient delivery networks optimized for urban landscapes, and the increasing utilization of smaller, more maneuverable vehicles ideal for navigating tight city streets. This is where innovative approaches, like consolidation centers located strategically to feed smaller vehicles, become essential.

Successful collaboration requires trust and a shared goal. Trucking companies, rail operators, and other stakeholders must be willing to collaborate, share data, and work towards a common objective. This might involve developing standardized procedures, creating collaborative platforms for information sharing, and fostering a culture of mutual understanding and transparency. Trust, in the context of shared information, is crucial in mitigating risks and optimizing efficiency.

Furthermore, fostering strong relationships between different actors in the supply chain significantly increases operational efficiency. This could include joint investments in infrastructure improvements or the development of shared technology platforms. Open communication channels help identify and quickly resolve unforeseen issues, preventing minor problems from escalating into major disruptions. Building collaborative relationships requires a conscious effort to foster mutual understanding and trust among partners.

Technological advancements are revolutionizing the landscape of road freight collaboration. Real-time tracking allows for improved visibility and predictive analysis, enabling proactive adjustments to delivery schedules. Automated systems streamline paperwork and reduce human error, leading to greater accuracy and efficiency. These technological integrations enable improved monitoring and precise management of all delivery stages and their interdependencies, increasing the overall effectiveness of the operations.

Finally, effective road freight collaboration requires a holistic approach, considering all aspects of the process. It's not just about the physical transfer of goods but also the efficient management of information, the coordination of personnel, and the optimization of resources. Achieving this requires a clear understanding of the challenges and opportunities inherent in integrating road freight with other modes, along with a strategic commitment to cooperation and innovation. The benefits of such an approach extend beyond improved efficiency and reduced costs; it ultimately leads to a more sustainable and resilient transportation network. The future of intermodal transport hinges heavily on effective road freight integration.

4.1 Truck-Rail Transfers

Efficient truck-rail transfers are critical for seamless intermodal transport. Minimizing handling time and maximizing capacity at transfer points directly impacts overall transit times and operational costs. Effective planning, including optimized scheduling and efficient equipment utilization, is paramount. Consideration must be given to the type of freight being transferred – whether it's containers, bulk goods, or specialized cargo – as different handling techniques are required.

The design of the transfer facility itself is crucial. Adequate space for maneuvering trucks and rail cars is essential, along with efficient loading and unloading ramps. The integration of automated systems, such as automated guided vehicles (AGVs) or automated stacking cranes, can dramatically speed up the transfer process. Furthermore, the facility must be strategically located to minimize transportation distances to and from major transportation hubs, effectively positioning it as a crucial link in the supply chain. Investing in robust infrastructure is undoubtedly a key factor in achieving operational excellence.

Staff training is equally important. Workers need comprehensive training in safe handling procedures, equipment operation, and efficient workflow processes. Regular training sessions and refresher courses ensure staff remains well-versed in industry best practices and able to respond effectively to unexpected situations. A well-trained workforce translates into minimized delays, reduced accidents, and a significant improvement in overall efficiency. Regularly assess and upgrade training programs to reflect technological advancements and evolving safety standards.

Choosing the right equipment is also vital for successful truck-rail transfers. This includes selecting appropriate trailers designed for easy loading and unloading onto rail cars, as well as choosing compatible rail cars themselves. Regular maintenance of this equipment is crucial to preventing breakdowns and delays. Investing in durable and reliable equipment that meets industry standards is a long-term cost-effective strategy. Regular inspections and preventative maintenance not only extend the lifespan of equipment but also mitigate potential operational disruptions.

Collaboration between trucking companies and railway operators is essential. Shared scheduling, data exchange, and clear communication protocols are all crucial for smooth operations. This collaboration should extend to the development of standardized procedures for transferring cargo, minimizing any ambiguity or potential for miscommunication. Developing strong working relationships and sharing real-time data fosters efficiency and transparency, leading to enhanced coordination between these modes of transport. Regular joint meetings and open communication channels are vital in addressing challenges and finding mutually beneficial solutions.

Real-time tracking and monitoring systems offer significant advantages in managing truck-rail transfers. These systems provide real-time visibility into the location and status of shipments, facilitating efficient scheduling and resolving potential delays proactively. Data analysis from these systems can be used to identify bottlenecks and optimize workflows, resulting in significant efficiency gains. Investing in robust tracking technologies is an investment in optimized operational management and improved overall performance.

Technological advancements such as the use of radio frequency identification (RFID) tags and advanced logistics software can significantly enhance efficiency. RFID technology allows for real-time tracking of individual units throughout the transfer process, enabling faster inventory management and improved accuracy. Advanced logistics software can automate many aspects

of the transfer process, optimizing scheduling, routing, and resource allocation. The integration of such technologies into the operational framework is key to reaching a higher level of efficiency.
Security measures are equally vital. Implementing robust security protocols, including access controls and surveillance systems, protects cargo from theft or damage during the transfer process. These measures, coupled with thorough staff training in security procedures, help minimize risks and maintain the integrity of the supply chain. A proactive security approach instills confidence among stakeholders and ensures the secure and efficient handling of valuable goods during transit.
Last but not least, environmental considerations are increasingly important. Adopting environmentally friendly practices, such as using low-emission vehicles and optimizing routes to reduce fuel consumption, helps to minimize the environmental impact of truck-rail transfers. Investing in sustainable solutions reflects a commitment to environmental responsibility and aligns with growing industry demands for eco-conscious operations. This proactive approach not only reduces environmental impact but can also contribute to enhancing a company's overall image.

4.2 Last-Mile Delivery

The last mile, that final leg of the journey, often presents the most significant logistical hurdles in intermodal transport. Unlike the seamless transfer between train and ship, the transition from a large-scale transport network to individual customer deliveries demands a fundamentally different approach. Efficiency here isn't simply about speed; it's about navigating the complexities of urban environments, varying delivery windows, and customer expectations. This necessitates a multi-pronged strategy that goes beyond simply selecting the right vehicle.

Consider the sheer volume of packages. The number of parcels needing delivery in major cities dwarfs the capacity of even the most efficient truck fleets. This creates congestion, delays, and increased costs. Effective last-mile solutions require strategic planning and innovation to optimize routes, consolidate shipments, and utilize available resources effectively. Think about utilizing micro-fulfillment centers strategically placed near high-density residential or commercial areas, or employing crowd-sourced delivery models leveraging personal vehicles. Technological innovations, too, are vital.

Real-time tracking and routing systems aren't merely for customer convenience; they're crucial for optimizing delivery efficiency. Accurate data on traffic patterns, delivery windows, and even weather conditions allows for dynamic route adjustments, ensuring faster delivery times and reduced fuel consumption. Advanced planning systems, integrating data from multiple sources, can anticipate demand fluctuations and proactively allocate resources, preventing bottlenecks and ensuring timely delivery even during peak periods. But technology alone is insufficient.

Successful last-mile delivery demands a deep understanding of local conditions. Understanding the unique characteristics of each delivery area – its traffic patterns, building access, and even local customs – is critical for effective planning. In densely populated urban areas, alternative delivery methods such as bicycle couriers, cargo bikes, or even drones become viable, environmentally friendly, and potentially cost-effective alternatives to traditional delivery trucks, significantly reducing carbon emissions and addressing traffic congestion. Smaller, more agile vehicles are ideal for navigating narrow streets and congested areas.

Furthermore, efficient last-mile delivery hinges on collaborative partnerships. Integrating with local businesses, leveraging their existing infrastructure and networks, can significantly reduce operational costs and improve overall efficiency. Collaboration extends beyond just physical infrastructure; it includes sharing real-time data and coordinating delivery schedules to minimize conflicts and maximize utilization of resources, enhancing overall delivery efficiencies and reducing environmental impact. This could involve partnering with local retailers to allow for package pick-up locations, or with smaller, localized delivery services that already have a strong presence in the community.

Finally, the human element remains indispensable. While technology optimizes efficiency, the personal touch of skilled delivery personnel remains crucial. Efficient delivery drivers are not merely package handlers; they are ambassadors of the brand, representing the company's image and directly impacting customer satisfaction. Proper training, clear communication channels, and investment in employee well-being contribute to higher morale, reduced turnover, and ultimately, enhanced customer experience. This results in greater efficiency and customer satisfaction, creating a positive feedback loop.

Last-mile delivery is not simply a logistical problem; it's a strategic challenge demanding innovative solutions and collaborative partnerships. It requires a holistic approach that considers technology, infrastructure, local conditions, and the human element to achieve optimal efficiency, reduce costs, and enhance customer satisfaction. This holistic approach, incorporating smart technologies and collaborative partnerships, is the key to unlocking the full potential of intermodal transport and ensuring the timely, efficient and cost effective delivery of goods to their final destination. By addressing these key considerations, businesses can significantly improve their last-mile operations and gain a competitive edge in the ever-evolving landscape of logistics.

5 Air Cargo Integration

Efficient airport logistics are paramount. Ground handling, encompassing activities from aircraft unloading to customs clearance and final delivery, significantly impacts overall transit time and cost. Streamlining these processes, perhaps through implementing advanced technology like automated guided vehicles (AGVs) for cargo movement within the airport, is crucial. Investing in robust infrastructure, including sufficient warehousing space and efficient sorting facilities, is equally vital to handle the growing volume of air cargo. Consideration should also be given to real-time tracking systems, enabling precise monitoring of each shipment's journey, providing vital information for both shippers and carriers, enhancing transparency and accountability. Finally, collaborative partnerships between airlines, ground handlers, and freight forwarders are essential for optimizing the entire ground handling process.

Air-sea intermodalism presents a fascinating blend of speed and cost-effectiveness. Combining the swiftness of air freight for long-distance transport with the economical nature of sea freight for the initial leg or less time-sensitive portions of the journey offers a compelling solution. Careful route planning is critical to leverage the strengths of each mode effectively. For example, high-value, time-sensitive goods may travel via air for a portion of the journey, potentially connecting to a sea route to reach a less accessible destination. This strategy requires sophisticated coordination and robust communication networks to ensure smooth transitions between modes. Furthermore, precise scheduling and efficient cargo handling at transfer points are essential to avoid delays and maintain the integrity of the cold chain for perishable goods.

Cargo consolidation is more than just stacking boxes; it's a strategic optimization process. Combining multiple smaller shipments into larger, more efficient units reduces the overall transportation cost per unit. This strategy is particularly beneficial for less-than-container-load (LCL) shipments, maximizing the utilization of available space on aircraft and reducing overall operational costs. However, careful consideration is needed in selecting appropriate consolidation points, optimizing container size, and employing advanced loading techniques to ensure the safety and secure delivery of individual shipments. Efficient labeling, clear documentation, and real-time tracking at each stage of the consolidation process help maintain transparency and accountability. Moreover, leveraging advanced planning and scheduling algorithms facilitates proactive identification of potential bottlenecks. Successful consolidation requires technology-based solutions and collaborative partnerships.

The efficient integration of air cargo into intermodal networks is not simply about moving goods; it's about optimizing the entire supply chain.

This involves close collaboration between various stakeholders – airlines, ground handlers, freight forwarders, customs agencies, and even shippers themselves. Effective information sharing, facilitated by advanced digital tracking systems, is critical. Transparency throughout the entire process enhances communication and allows for proactive problem-solving. For example, real-time updates on flight schedules and cargo status enable all stakeholders to adapt quickly to any unexpected disruptions, mitigating potential delays. Furthermore, the development of standardized procedures and protocols for air cargo handling across different airports and countries facilitates seamless intermodal transfer and reduces the risk of errors or delays. The ultimate goal is to create an integrated system that is both efficient and reliable. To ensure the long-term viability of this integrated system, continuous improvement is essential, relying on data analysis, feedback mechanisms, and technological advancements to enhance overall efficiency and reduce costs. The future of air cargo integration hinges on a commitment to innovation and collaborative partnerships.

5.1 Airport Logistics

Efficient airport logistics are crucial for seamless integration within the broader intermodal transport network. The speed and global reach of air freight necessitate optimized ground handling procedures to avoid delays and maintain the competitive advantage of air transport. This involves a complex interplay of various stakeholders, including airlines, ground handling agents, customs authorities, and freight forwarders. Streamlining these interactions is paramount.

Effective airport logistics begin with careful planning and design of airport infrastructure. Dedicated cargo terminals, with sufficient space for aircraft parking, loading docks, and warehousing, are essential. The layout should minimize the distance cargo travels between the aircraft and the final destination, whether it's another mode of transport or direct delivery. Strategic placement of screening and customs facilities also minimizes congestion and delays. Efficient infrastructure is the bedrock upon which faster and more efficient logistics stand.

Technological advancements play a pivotal role in modernizing airport logistics. Real-time tracking systems allow for constant monitoring of cargo throughout its journey, from arrival at the airport to final delivery. This visibility improves efficiency and reduces the risk of delays or losses. Advanced baggage handling systems automate the sorting and transportation of luggage, minimizing manual handling and accelerating the baggage claim process. Integration of these systems with other modes of transport is key for seamless transfer.

Data-driven decision-making is increasingly important in airport logistics. Analytics can be used to predict peak periods, optimize staffing levels, and improve resource allocation. This predictive capability allows airport operators to proactively address potential bottlenecks and ensure smooth operations, even during periods of high volume. By anticipating challenges, airports are better equipped to handle them effectively. The ability to utilize these systems is also crucial for better forecasting capabilities, and it helps to prepare for potential challenges.

Security is a non-negotiable aspect of airport logistics. Stringent screening procedures are crucial for both passenger and cargo safety. Effective collaboration between airport security personnel, customs officials, and other agencies is necessary to ensure a high level of security while minimizing disruptions. Layered security approaches, combined with robust technology, such as advanced imaging systems, can enhance both safety and efficiency. Security is not just about preventing malicious actions, it is also about maintaining trust in the system. The perception of security is as important as the reality.

Collaboration among stakeholders is essential for success. Open communication channels between airlines, ground handlers, freight forwarders, and other parties are crucial for effective coordination and problem-solving. Establishing clear communication protocols and shared information systems can significantly improve efficiency and reduce the likelihood of conflicts or delays. This includes agreements on procedures for handling delays, emergencies, and conflicts of interest. Effective communication also extends to informing customers about their shipments and resolving concerns.

Sustainable practices are also increasingly important in airport logistics. Reducing carbon emissions through efficient operations and the use of sustainable technologies, such as electric ground support equipment, is crucial for minimizing the environmental impact of air freight. This aligns with broader efforts to make transportation globally more environmentally friendly, and it contributes to the overall sustainability goals of the industry. Choosing environmentally friendly options for ground transportation also is crucial in minimizing the negative footprint of airport operations. Finally, effective workforce management is a key component of airport logistics. Training employees on the latest technologies and procedures is crucial for maintaining a highly skilled and efficient workforce. Employee engagement and retention strategies are essential to avoid staff shortages, which can lead to delays and inefficiencies. A well-trained and satisfied staff significantly contributes to smooth operations and positive customer experience. Employee wellbeing and development are key investments in efficient airport logistics.

The successful integration of air cargo into intermodal systems requires a holistic approach to airport logistics, encompassing infrastructure, technology, security, collaboration, sustainability, and workforce management. By focusing on these areas, airports can optimize their operations, contributing to the overall efficiency and competitiveness of the global intermodal transport network. Attention to detail and a focus on continuous improvement are key to achieving excellence in airport logistics. The aim is not merely to move goods, but to move them efficiently, safely, and sustainably.

5.2 Air-Sea Intermodalism

Air-sea intermodalism offers a compelling blend of speed and economy, making it a crucial component of global supply chains. Imagine a shipment of high-value electronics originating in Shenzhen, China, destined for a major European city. Sea freight, while economical, takes weeks. Air freight, conversely, is rapid but significantly more expensive. The optimal solution? Often, it's a combination of both. This involves transporting the goods by sea to a strategically located port, typically a major hub with robust air cargo facilities, and then swiftly transferring the shipment to an aircraft for the final leg of the journey. This minimizes costs while still ensuring timely delivery.

Effective air-sea intermodalism demands seamless coordination between various stakeholders. This includes the shipper, the ocean carrier, the air freight forwarder, and the ground handling agents at both the seaport and the airport. Clear communication, standardized procedures, and robust tracking systems are paramount to avoid delays and ensure efficient cargo flow. Delays can easily snowball, impacting delivery schedules and potentially harming the reputation of all parties involved. Precise timing is critical, as the success of this hybrid transport method hinges on the synchronization of different transit modes.

Efficient port infrastructure is undeniably critical for successful air-sea intermodal operations. The port must have sufficient capacity to handle both large sea vessels and the volume of air cargo that will be transferred. Modern container handling equipment is essential for quick and efficient cargo transfer between ships and trucks, along with dedicated facilities for customs clearance and cargo inspection. Insufficient infrastructure leads to congestion and delays, negating the benefits of this strategy. The closer the airport is to the port, the better, minimizing the time and cost associated with ground transportation. Direct access to well-maintained roadways connecting the port and airport is also a vital consideration.

Choosing the right sea and air routes is equally important. Optimizing the sea leg often prioritizes cost-effectiveness, selecting a route with lower freight rates, even if it slightly increases transit time. The air leg, on the other hand, prioritizes speed and reliability, utilizing established routes with a proven track record of punctuality. Advanced route planning software incorporating real-time data on weather, traffic, and airport capacity helps optimize both legs, creating the most efficient overall journey. This proactive approach reduces unforeseen disruptions that often plague less planned operations.

Cargo consolidation is another key element. Consolidating multiple shipments from various shippers into larger, more efficient units for both sea and air transit maximizes utilization of available space and reduces per-unit transportation costs. This requires sophisticated logistics planning and management, frequently handled by specialized freight forwarders with the expertise to coordinate the intricacies involved. The optimal size of consolidated shipments is carefully determined based on the carrying capacities of both sea vessels and aircraft, aiming to avoid wasted space and minimize shipping costs.

Security throughout the entire process is a crucial concern. Air-sea intermodal shipments often carry high-value goods, making them attractive targets for theft or damage. Robust security measures are needed at each stage, from the initial packaging and loading at the origin to final delivery at the destination. This involves the implementation of advanced tracking systems to monitor cargo location and status in real-time, alongside thorough security checks and risk assessments to mitigate potential vulnerabilities throughout the supply chain. Investing in robust security systems demonstrates a commitment to safeguarding customer goods and maintaining a high level of integrity within the industry.

The future of air-sea intermodalism lies in increased automation and data-driven optimization. Technologies like blockchain, offering enhanced security and transparency across the entire supply chain, are becoming increasingly important. The development of more accurate predictive analytics models enables better forecasting of demand and optimizing routes and schedules to minimize delays. Autonomous vehicles and drones could revolutionize last-mile delivery, further enhancing efficiency and reducing costs. Continuous technological advancements will continue to redefine the way air-sea intermodal transport operates. The continued success of this approach depends on adapting and implementing these innovations, staying ahead of the curve in the ever-evolving landscape of global logistics.

5.3 Cargo Consolidation

Consolidating cargo, at its core, is about optimizing shipment size and reducing overall transportation costs. Think of it like a game of Tetris, but with far more valuable and diverse pieces. By grouping smaller shipments together, you create larger, more efficient units, better utilizing available space within containers, trucks, or aircraft. This not only minimizes the number of individual shipments—reducing handling and paperwork—but also significantly lowers per-unit transportation expenses. This principle applies equally whether you're dealing with a handful of small packages bound for a single destination or a vast array of goods destined for a global network of recipients.

Effective consolidation requires meticulous planning and coordination. You need a comprehensive understanding of your shipment profile, including the volume, weight, dimensions, and destination of each individual item. Sophisticated software can be your best friend here, helping you analyze your data to identify optimal grouping strategies. This might involve sorting shipments by destination, size, or even fragility, ensuring that compatible goods are bundled together, minimizing the risk of damage during transit. This is where your expertise becomes truly invaluable; understanding product specifications is critical for successful consolidation. Careful consideration of weight distribution, for example, is paramount to preventing imbalances and potential structural failure within the consolidated units.

Beyond the technical aspects, building strong relationships with your logistics partners is vital. Open communication is key; regular updates between shippers, consolidators, and carriers are essential to ensure everyone is on the same page, from initial planning through final delivery. This proactive approach helps anticipate and resolve potential issues promptly, minimizing disruptions and delays. Remember, transparency and trust are crucial for building resilient partnerships within the supply chain. The ability to react to unforeseen events, such as port congestion or weather delays, is crucial to maintaining efficiency and meeting deadlines.

While the benefits of cargo consolidation are numerous, it's not without its challenges. Balancing the desire for cost-effectiveness with the need for timely delivery requires skillful navigation. Improper consolidation can lead to delays, damage, or even lost goods. Understanding each mode of transport's specific requirements and limitations—weight restrictions on aircraft, for instance, or space constraints in ocean containers—is crucial for successful implementation. Furthermore, strict adherence to safety regulations, especially regarding hazardous materials, is non-negotiable. Failure to comply can result in significant penalties and reputational damage. Safety protocols must always be prioritized in the consolidation process.

Investing in advanced technology can drastically improve your consolidation strategy. Real-time tracking systems offer unparalleled visibility into your shipments, allowing you to monitor their progress and react to potential problems proactively. Predictive analytics can even forecast potential bottlenecks or disruptions, enabling you to make adjustments ahead of time. These technological advancements don't just improve efficiency; they enhance security and transparency, ultimately building trust with your clients and partners. Modern technologies offer an advantage that can't be ignored. This level of technological sophistication provides a competitive edge in the ever-evolving landscape of intermodal transportation.

Consider the potential implications for your bottom line. By implementing a well-designed consolidation strategy, you can streamline your operations, reduce costs, and improve the overall efficiency of your supply chain. The advantages extend beyond cost savings; enhanced speed and reliability lead to improved customer satisfaction and a stronger competitive position in the marketplace. It's about optimizing every aspect of your shipment process, from origin to destination. You're not just moving goods; you're optimizing a complex network of processes and relationships. This comprehensive view is key to success.

The long-term success of your consolidation strategy hinges on continuous improvement and adaptation. Regularly review your processes, analyze your data, and identify areas for optimization. The shipping world is dynamic; staying ahead of the curve through continuous learning and adaptation is essential. The industry is marked by constant evolution; embracing new technologies and strategies ensures long-term competitiveness and success. Therefore, making continuous improvements is not optional but rather essential for ongoing success within this industry. This necessitates a commitment to ongoing professional development and a proactive approach to staying informed about industry best practices. This proactive approach is critical in remaining ahead of the curve in the rapidly evolving field of global logistics.

6 Technological Advancements

Digital tracking systems offer unprecedented visibility into the movement of goods across various modes. Imagine a single platform displaying the real-time location of a container, from its origin at a manufacturing plant in China, through its journey by sea to a port in Rotterdam, its subsequent transfer to a rail terminal, and finally, its delivery by truck to a warehouse in London. This end-to-end tracking capability significantly improves efficiency by reducing delays, streamlining communication between stakeholders, and providing crucial data for predictive analysis. Real-time updates empower quicker decision-making, leading to minimized disruptions and improved overall supply chain responsiveness. The adoption of globally standardized data protocols is crucial for the seamless integration of disparate systems and maximizing the utility of these tracking technologies.

Predictive analytics leverages the vast amount of data generated by tracking systems and other sources to forecast potential bottlenecks and optimize logistical operations proactively. By analyzing historical data, weather patterns, traffic conditions, and other relevant factors, sophisticated algorithms can predict delays, optimize routes, and recommend ideal schedules for greater efficiency and cost savings. For instance, a predictive model might anticipate a port congestion and suggest rerouting cargo through an alternative port, avoiding substantial delays and financial losses. This proactive approach goes beyond mere reaction to problems; it allows for preventative measures that ensure smooth, consistent, and economical operations. The accuracy and effectiveness of predictive analytics depend heavily on the quality and completeness of the data fed into the models, highlighting the critical role of robust data collection and management systems.

Blockchain technology, with its decentralized and tamper-proof nature, presents a revolutionary opportunity for enhancing security and transparency in intermodal transport. By recording every transaction and transfer of goods on a shared, immutable ledger, blockchain eliminates the risk of fraud and data manipulation, promoting trust between all parties involved. This secure and transparent system can track ownership and provenance of goods, facilitating seamless customs clearance and reducing documentation delays, creating a more efficient and reliable system for handling international shipments. The implementation of blockchain in intermodal logistics is still nascent, but its potential to revolutionize the industry is immense, paving the way for greater efficiency, security, and trust.

Automation is rapidly transforming many aspects of intermodal operations. From automated container handling systems in ports to robotic sorting in rail yards, automation improves efficiency, reduces labor costs, and minimizes the risk of human error. Automated guided vehicles (AGVs) and automated stacking cranes are just a few examples of technology streamlining operations. The adoption of automation technologies necessitates significant upfront investment, but the long-term benefits, in terms of productivity and efficiency gains, are substantial. The integration of automation necessitates careful planning and consideration of workforce adaptation strategies to ensure a smooth transition and minimize disruptions.

Data-driven optimization involves leveraging the insights derived from vast amounts of data to continually refine and improve intermodal processes. By systematically analyzing operational data, businesses can identify areas for improvement, refine scheduling algorithms, and optimize resource allocation. This iterative process of analysis, refinement, and re-optimization leads to a continual increase in efficiency and cost-effectiveness. This includes sophisticated route optimization, more effective vehicle scheduling, and predictive maintenance programs to reduce downtime. Investing in robust data analytics capabilities and skilled data scientists is essential for harnessing the full potential of data-driven optimization. The ability to adapt quickly to changing conditions and to continuously refine operations based on real-time feedback is key to achieving lasting improvements.

6.1 Digital Tracking Systems

Real-time tracking is paramount in today's interconnected global supply chains. Imagine a scenario: a shipment of vital medical supplies is delayed, causing a shortage in a remote hospital. This isn't hypothetical; such incidents highlight the critical need for precise, readily available information. Digital tracking systems offer that precision, providing end-to-end visibility into the movement of goods across multiple transportation modes. These systems use a combination of technologies, including GPS, RFID, and IoT sensors, to monitor the location, condition, and handling of shipments. This level of granularity allows for proactive intervention, minimizing disruptions and ensuring timely delivery.

For beginners, the integration of these systems isn't always intuitive. Starting with a phased approach is crucial. First, identify the critical points in your intermodal network that require the most stringent tracking. Perhaps it's the transfer between rail and road, a point historically known for potential delays or mishandling. Prioritize those areas for initial system implementation. This targeted strategy minimizes the upfront investment and ensures a quick return on your investment by focusing on the highest-impact areas first. Don't try to overhaul everything at once; a gradual integration allows for smoother adjustments and better user adoption.

Choosing the right technology is also essential. GPS provides broad location data, while RFID tags offer more detailed information about individual units, particularly useful when handling large quantities of similar goods. Consider the specific needs of your intermodal operations. Will you be tracking individual containers, pallets, or even individual items? The answer guides your technology selection. Remember, robust data security is crucial. Choose a system that adheres to the highest data protection standards, encrypting your sensitive shipment data to prevent unauthorized access or breaches. This protects your valuable cargo information and your company's reputation.

The practical application of these tracking systems extends beyond simple location monitoring. Many modern systems integrate with advanced analytics platforms, generating reports that analyze shipment trends, identify bottlenecks, and predict potential delays. This data-driven approach enables proactive adjustments to scheduling and routing, optimizing the entire logistics chain and increasing efficiency. For example, the system might highlight a consistently congested rail yard, suggesting alternative routes or optimizing loading schedules to minimize delays at that specific point. The information empowers you to make informed decisions, improving overall operational efficiency and customer satisfaction.

Beyond immediate operational benefits, digital tracking systems significantly enhance transparency and accountability throughout the supply chain. All stakeholders – from shippers to carriers to receivers – can access real-time information about a shipment's progress, fostering greater trust and collaboration. This transparency is particularly important in complex intermodal operations, where multiple parties are involved. Having a centralized system for tracking and sharing information eliminates confusion and streamlines communication. Furthermore, this transparency often leads to better relationships between partners and the potential for cost-sharing agreements and improved service-level agreements.

Implementing a digital tracking system isn't simply a technical upgrade;

it's a strategic decision that significantly impacts your business operations. Start small, focus on high-impact areas, and carefully choose technologies to meet your specific needs. Always prioritize data security and remember that the benefits extend beyond just tracking – enhanced analytics, improved transparency, and stronger stakeholder relationships are all significant advantages that contribute to a more efficient and resilient intermodal operation in the long run. Don't underestimate the value of this investment; it is a foundation for sustainable growth and success in the competitive world of global logistics. The initial investment will likely pay off in improved efficiency, reduced costs, and enhanced customer satisfaction. Treat this implementation as a long-term project, integrating training and ongoing support into your strategy for maximum impact. Regular reviews will be crucial to measure success and identify areas for optimization.

6.2 Predictive Analytics

Predictive analytics offers a powerful tool for optimizing intermodal transportation networks. By leveraging historical data, current conditions, and sophisticated algorithms, businesses can anticipate disruptions, optimize routes, and improve overall efficiency. This leads to significant cost savings and enhanced customer satisfaction. Imagine the possibilities: predicting potential delays due to weather patterns, optimizing container stacking in ports based on expected arrival times, or even dynamically adjusting shipping routes in real-time to avoid congestion. The potential for optimizing the entire intermodal supply chain is immense, leading to significant cost reductions and streamlined operations.

Forecasting demand is another crucial application. Accurate predictions allow companies to better allocate resources, ensuring sufficient capacity to meet anticipated demand while avoiding unnecessary overspending. This is particularly relevant in the volatile world of global trade, where fluctuations in demand are common. Historical data on seasonal trends, economic indicators, and even social media sentiment can be incorporated into predictive models to provide highly accurate forecasts. This data-driven approach allows for proactive capacity planning, preventing costly shortages or surpluses.

But predictive analytics isn't just about forecasting; it's also about risk management. Identifying potential bottlenecks, delays, or security threats before they occur allows for proactive mitigation strategies. Analyzing historical data on incidents like port congestion, train derailments, or even inclement weather can reveal patterns and probabilities, enabling companies to implement contingency plans and minimize disruptions. Such early warnings drastically reduce potential losses and ensure greater operational resilience.

The application of machine learning within predictive analytics further enhances its capabilities. Machine learning algorithms, trained on extensive datasets, can identify complex patterns and relationships that might be missed by traditional statistical methods. These algorithms can continuously learn and improve their predictive accuracy over time. This ability to adapt to changing conditions and uncover hidden insights is a game-changer, enabling more accurate and effective decision-making. This also applies to predicting maintenance needs for equipment, allowing for scheduled maintenance to reduce unexpected downtime and operational costs.

Implementing predictive analytics successfully requires a strategic approach. Start by identifying key performance indicators (KPIs) relevant to your business. This might include on-time delivery rates, transportation costs, or dwell time in ports. Then, gather and clean the necessary data, ensuring its accuracy and completeness. This stage requires careful planning and collaboration with various stakeholders across the supply chain. Consider the sources of your data, ensuring consistent quality. Inaccurate data will lead to poor predictions, negating the value of the analytics.

Choosing the right predictive modeling techniques is also critical. The complexity of the model should match the complexity of the problem being addressed. Simple linear regression may suffice for some tasks, while more advanced algorithms such as neural networks might be necessary for others. The expertise to design and implement these models is key, and might require outsourcing to specialists who can assist in creating predictive models that are practical for your application. These specialists should have a strong understanding of intermodal transportation systems to appropriately interpret the results.

Finally, successful implementation demands a culture of data-driven decision-making. The insights gained from predictive analytics should be integrated into existing workflows and used to inform strategic and operational decisions. This requires clear communication and collaboration between IT teams, operations managers, and executives. Real-time dashboards and reporting tools can facilitate effective communication and enable stakeholders to monitor performance and make data-informed decisions quickly. It's not enough to simply generate predictions; companies need to act upon them strategically to benefit. This involves not only technological capabilities but changes in the company's organizational structure to fully exploit the results of the analysis.

Therefore, predictive analytics represents a significant advancement for the intermodal transportation industry. It provides a pathway towards greater efficiency, reduced costs, and improved resilience to unforeseen circumstances. By leveraging the power of data and advanced analytical techniques, businesses can navigate the complexities of global supply chains and optimize their operations for maximum success in a dynamic and competitive market. Investing in robust systems and building a strong understanding of these analytical methods will be essential to harnessing these benefits. The future of intermodal transport is deeply intertwined with the effective utilization of predictive analytics; those who adapt early will be best positioned for long-term success.

6.3 Blockchain Technology

Blockchain technology offers a revolutionary approach to enhancing transparency and security within intermodal transportation networks. Imagine a system where every step of a shipment's journey, from origin to final delivery, is immutably recorded on a shared, secure ledger. This is the power of blockchain. Its decentralized nature eliminates the need for a central authority, reducing the risk of data manipulation or fraud. Instead, multiple participants—shippers, carriers, customs agents, and more—can access verified information simultaneously, fostering greater trust and collaboration throughout the supply chain. This real-time visibility not only improves efficiency but also significantly minimizes delays and disputes.

The benefits extend beyond simple tracking. Smart contracts, self-executing agreements written into the blockchain, can automate payments upon the successful completion of predetermined milestones. This eliminates paperwork, accelerates settlements, and reduces costs associated with manual reconciliation. Consider the potential for automating customs clearance, significantly reducing the time goods spend waiting at borders. Blockchain can also integrate with other technologies, such as IoT sensors, to provide real-time updates on cargo location, temperature, and condition. This granular level of data facilitates proactive issue resolution and minimizes potential losses due to spoilage or damage. For instance, if a refrigerated container's temperature deviates from the set point, an immediate alert can be sent to relevant parties, preventing costly product loss.

Furthermore, the inherent security of blockchain significantly reduces the risk of cargo theft and counterfeiting. The immutability of the data ensures that any attempts to tamper with records are immediately detectable, providing a strong deterrent against fraudulent activities. This increased security can lead to lower insurance premiums and reduced losses from theft or damage. We're talking about a potentially massive reduction in overall operational costs. Companies can redirect the money saved into infrastructure improvements or investments in new technologies. This positive feedback loop could accelerate innovation and modernization across the intermodal sector.

One particularly interesting application lies in the creation of digital bills of lading (B/Ls). Currently, paper-based B/Ls are prone to fraud, loss, and delays. A blockchain-based B/L, however, offers an immutable record of ownership and transfer, reducing disputes and accelerating customs clearance. This digitalization process not only streamlines documentation but also enhances the overall efficiency of global trade. Imagine eliminating the need for multiple parties to handle and verify physical documents, instead, all relevant information is instantly accessible to all authorized parties. This level of transparency will lead to significant savings in time and resources. However, it's crucial to acknowledge the challenges associated with blockchain implementation in the complex world of intermodal transport. Integration with existing legacy systems can be technically challenging and costly. The need for widespread adoption across the industry is also paramount for blockchain to achieve its full potential. Addressing scalability issues and ensuring interoperability between different blockchain platforms are also key concerns that need to be tackled proactively. Educational initiatives and clear regulatory frameworks are vital for ensuring a successful and widespread transition. For beginners, understanding these technical and logistical hurdles is critical.

Despite these challenges, the potential benefits of blockchain are too significant to ignore. It's a transformative technology with the power to reshape the entire intermodal transportation landscape. Its ability to improve transparency, security, and efficiency makes it a crucial technology for anyone involved in global logistics. As the technology matures and adoption increases, we can expect to see even more innovative applications emerge, further optimizing the flow of goods across the globe and potentially revolutionizing the way we think about global trade. The initial investment may seem daunting, but the long-term benefits in terms of cost savings, increased security, and improved efficiency far outweigh the initial hurdles. Embracing this technology is not just about keeping up with the times; it's about strategically positioning oneself for future success in a rapidly evolving industry. The future of intermodal transport is undoubtedly intertwined with the continued development and implementation of blockchain technology.

6.4 Automation in Logistics

Automated guided vehicles (AGVs) are transforming warehouse operations, smoothly navigating complex layouts to transport goods with remarkable precision. Their consistent speed and efficiency far surpass human capabilities, significantly reducing transit times and optimizing space utilization within the facility. This translates directly to lower operational costs and enhanced delivery performance, making them a crucial investment for businesses aiming for a competitive edge in today's fast-paced market. The integration of AGVs requires careful planning and implementation, encompassing system compatibility, infrastructure modifications, and employee training, but the long-term benefits are undeniable.

Implementing robotic process automation (RPA) in logistics goes beyond simply automating repetitive tasks. These systems can handle complex processes like customs declarations, freight booking, and invoice processing, leading to fewer human errors and substantial time savings. By taking over these administrative duties, RPA frees up human workers to focus on more strategic activities, thus improving overall team productivity and efficiency. However, successful RPA implementation necessitates a detailed analysis of current workflows and choosing the right software tailored to your specific needs and business scale. Ignoring this can result in a less than optimal integration process.

Automated sorting systems are revolutionizing the handling of vast quantities of packages, particularly in large distribution centers. These sophisticated systems utilize advanced technologies like optical character recognition (OCR) and barcode scanning to rapidly sort packages based on various criteria – destination, size, priority – and route them directly to the appropriate loading docks or delivery vehicles. This significantly speeds up the sorting process, reduces the likelihood of mis-sorting and subsequent delays, contributing to the prompt and efficient delivery of packages. The initial investment in such systems can be substantial, but the return on investment, calculated over time based on productivity gains and reduced labor costs, is considerable.

The rise of automated container terminals is fundamentally altering maritime port operations. Remote-controlled cranes, automated stacking cranes, and sophisticated scheduling software now work in tandem to handle container shipments with unprecedented speed and efficiency. These advancements minimize manual labor, reduce congestion, and improve overall port throughput. Moreover, the adoption of these automated systems leads to safer working conditions for dockworkers, while reducing the risk of human error. This is not simply about replacing human workers; it's about enhancing the overall efficiency of the entire port operation, ultimately benefitting the global supply chain.

Predictive maintenance, powered by machine learning and data analytics, is becoming a standard practice in logistics. By analyzing data from various sources—sensor readings from vehicles and equipment, weather forecasts, traffic patterns— sophisticated algorithms can predict potential equipment failures before they occur. This allows for proactive maintenance, reducing downtime, preventing costly repairs, and ensuring the uninterrupted flow of goods. It is a shift from reactive to proactive management, a critical step towards achieving greater operational reliability and efficiency within logistical systems. The implementation requires careful data collection, analysis, and the integration of appropriate software solutions that can interpret this data and deliver actionable insights.

Digital twin technology offers a powerful tool for logistics optimization. This involves creating a virtual replica of a physical asset or process, allowing for the simulation of various scenarios before implementing them in the real world. This helps to identify potential bottlenecks, optimize routes, and improve the overall efficiency of the supply chain. For example, a digital twin of a warehouse can help optimize the layout and flow of goods within the facility or the simulation of different transportation routes can help determine the most efficient options. The benefits lie in the ability to test multiple strategies and make informed decisions, minimizing risk and maximizing resource utilization before any physical changes are made. However, accurate modeling and reliable data are crucial for achieving effective results.

Drone technology is rapidly gaining traction for last-mile delivery, offering a viable solution for delivering smaller packages to remote or congested areas more efficiently. Drones can bypass traffic congestion, deliver packages directly to their destinations with greater speed and precision compared to traditional delivery methods. This is particularly beneficial in urban environments, where access limitations and heavy traffic often cause significant delivery delays. While regulatory hurdles and technological refinements remain, the potential for improved speed, efficiency, and reduced costs associated with drone delivery is significant and promises to reshape last-mile logistics.

The integration of these technologies is not simply a matter of replacing human workers with machines; it's about creating a synergistic relationship where human expertise and machine efficiency work together. Humans provide the strategic thinking, problem-solving, and adaptability that machines currently lack, while automation handles the repetitive, data-intensive tasks, enhancing efficiency, and overall effectiveness. A successful automation strategy requires a carefully considered approach, factoring in existing infrastructure, workforce skills, and the specific needs of the logistics operation. The ultimate goal is not just automation for the sake of it, but rather the optimization of the entire process to maximize efficiency and overall performance. The transition to a more automated logistical system will undoubtedly present challenges, but the rewards in terms of speed, efficiency, reduced costs and improved safety are ultimately worth the effort and planning required for successful implementation.

6.5 Data-Driven Optimization

Data analytics offer unparalleled opportunities to enhance efficiency and profitability within intermodal transportation. By meticulously collecting and analyzing data from various sources – encompassing everything from GPS trackers on individual containers to weather forecasts impacting schedules – we gain incredibly valuable insights into operational bottlenecks and potential improvements. This allows for proactive adjustments and optimization strategies, minimizing delays and maximizing resource utilization. Consider, for example, the real-time tracking of containers. Knowing the precise location and status of each shipment allows for predictive maintenance on transport vehicles, preventing costly breakdowns and delays. Furthermore, this data can be integrated with route optimization software, dynamically adjusting routes based on traffic patterns, weather conditions, and even anticipated demand fluctuations. The result is a more reliable and cost-effective transport network. Imagine the potential savings achieved by preventing even a single significant delay, considering the cascading effects on subsequent shipments and the associated costs.

But the benefits extend beyond mere efficiency. Data analytics can also reveal previously unseen patterns in demand and supply. By analyzing historical shipment data, alongside external factors such as seasonal trends and economic indicators, businesses can better predict future needs, optimizing inventory management and reducing warehousing costs. This proactive approach minimizes unnecessary storage and reduces the risk of stock-outs, leading to significant cost savings and improved customer satisfaction. Imagine the precise forecasting of peak shipping seasons, allowing for the proactive allocation of resources and the prevention of congestion at major intermodal hubs.

Moreover, the integration of data-driven optimization with other technological advancements, such as automation in logistics, exponentially increases its impact. Automated systems can process vast amounts of data significantly faster than humans, identifying subtle patterns that might otherwise be missed. Combined with machine learning algorithms, these systems can continuously refine their optimization strategies, adapting to changing conditions in real-time and providing increasingly accurate predictions. This continuous learning process ensures that optimization strategies are always evolving and adapting to the dynamic nature of the intermodal transport industry, guaranteeing sustained improvements.

However, the successful implementation of data-driven optimization requires a thoughtful and strategic approach. Simply collecting data isn't enough; the data needs to be accurate, reliable, and consistently collected. This necessitates the implementation of robust data collection systems and stringent quality control procedures. Furthermore, the data must be structured and organized in a way that is easily accessible and analyzable by the chosen software and algorithms. Investing in the right infrastructure and training the workforce to utilize the data effectively are critical steps in the successful implementation of a data-driven optimization strategy.

Beyond the technical aspects, organizational considerations are equally crucial. Effective data-driven optimization requires collaboration across different departments and even between different organizations involved in the intermodal transport chain. Sharing data and fostering transparency between stakeholders is essential to achieving the full benefits of this technology. Consider the sharing of data between trucking companies and railway operators, allowing for smoother transitions between different modes of transport and the minimization of delays at transfer points. This collaborative approach not only enhances efficiency but also strengthens relationships within the broader intermodal community.

Finally, while data-driven optimization offers immense potential for improving efficiency, profitability, and sustainability within intermodal transportation, it is vital to acknowledge its limitations. The accuracy of predictions relies on the quality and completeness of the data used. Outliers and unexpected events, such as natural disasters or geopolitical instability, can significantly impact the accuracy of forecasts. Therefore, it is essential to complement data-driven optimization with human oversight, ensuring that the system's recommendations are evaluated critically and adjusted as needed. This balanced approach—combining the power of data analytics with human judgment—maximizes the benefits and minimizes the risks associated with this transformative technology.

7 Future of Intermodal Transport

The increasing demand for efficient and sustainable global logistics necessitates a paradigm shift in intermodal transport. We're moving beyond simply connecting different modes; we're looking towards seamless, integrated systems that optimize every stage of the journey. This requires a holistic approach, encompassing technological innovation, sustainable practices, and a careful consideration of evolving global trade dynamics. The future isn't just about faster transit times, it's about smarter, greener, and more resilient networks.

Consider the implications of autonomous vehicles. Self-driving trucks, for example, could significantly reduce labor costs and improve efficiency in last-mile delivery, a notoriously challenging aspect of intermodal transport. Furthermore, their potential to operate 24/7 could drastically increase throughput at intermodal terminals and reduce congestion on roadways. However, integrating autonomous vehicles into existing infrastructure requires careful planning and substantial investment in supporting technologies and regulations. We must also address potential job displace-

ment issues and ensure safety protocols are robust and rigorously tested.
Beyond autonomous vehicles, drone technology presents another transformative opportunity. Imagine a future where smaller, time-sensitive packages are delivered directly from rail or maritime hubs, bypassing congested roads and reducing delivery times. This could revolutionize e-commerce logistics and create new business models within the intermodal sector. Yet, the integration of drones faces hurdles. Regulation regarding airspace management and safety remains a significant barrier. We also need to address issues of range, payload capacity, and the development of reliable and scalable drone infrastructure.

Sustainable solutions are no longer optional; they are essential for the long-term viability of intermodal transport. The industry must embrace greener fuels, optimize routes to reduce emissions, and invest in energy-efficient technologies across all modes. This includes exploring alternative fuels for ships and trains, designing more fuel-efficient vehicles, and investing in infrastructure that supports electric and hydrogen-powered transport. Embracing sustainability isn't just about environmental responsibility; it's also about enhancing the industry's long-term resilience and competitiveness. Regulations are shifting globally, and companies that fail to adapt risk losing market share and facing significant penalties.

Global trade patterns are constantly evolving, and the future of intermodal transport is intrinsically linked to these shifts. The rise of e-commerce has driven unprecedented demand for faster and more reliable delivery, forcing the industry to adapt to new consumer expectations. Emerging markets in Asia and Africa present significant growth opportunities but also pose logistical challenges, requiring the development of new infrastructure and efficient transport corridors. Geopolitical events and trade agreements continue to shape the global landscape, underscoring the importance of flexible and adaptable intermodal networks.

The convergence of various technologies will undoubtedly redefine the landscape of intermodal transport. The integration of blockchain technology, for instance, can enhance security and transparency throughout the supply chain, improving traceability and reducing the risk of fraud. Advanced data analytics will enable more accurate forecasting, allowing operators to optimize routes, schedules, and resource allocation, improving overall efficiency and reducing costs. The combination of these technologies, along with ongoing advancements in automation and robotics, promises a future of smarter, more efficient, and sustainable global logistics. However, realizing the full potential of these technologies requires collaboration between industry stakeholders, policymakers, and researchers, ensuring that innovation is implemented responsibly and sustainably.

Finally, effective workforce development is paramount. The shift toward automation and the adoption of new technologies will require a skilled workforce capable of operating and maintaining sophisticated equipment and managing complex systems. Investing in training and education programs will be crucial to bridging the skills gap and ensuring the industry can adapt to the evolving demands of the future. This includes fostering collaboration between educational institutions and industry leaders to ensure curricula reflect the ever-changing needs of the intermodal transport sector. A well-trained and adaptable workforce is the foundation upon which future success is built.

7.1 Sustainable Solutions

The environmental footprint of global intermodal transport is substantial. Increased fuel consumption, greenhouse gas emissions, and air and noise pollution are significant concerns. Mitigating these impacts requires a multi-pronged approach, focusing on technological innovation, operational efficiencies, and policy changes. We need to move beyond simply minimizing harm and embrace truly sustainable practices.

Shifting towards alternative fuels is crucial. Biofuels, hydrogen, and electricity offer viable, albeit still developing, alternatives to traditional fossil fuels. Investing in research and development, as well as the necessary infrastructure to support these fuels, will be essential for large-scale adoption. This includes not only the vehicles themselves, but also the charging stations, refueling stations, and energy grids needed to power a greener transportation network. Furthermore, careful consideration must be given to the life cycle assessment of these alternative fuels to ensure their overall sustainability. This means examining everything from production and transportation to their ultimate disposal or reuse.

Optimizing logistics and route planning is another key element of sustainable intermodal transport. Advanced technologies like AI-powered route optimization software can significantly reduce fuel consumption by identifying the most efficient routes, minimizing empty miles, and improving overall fleet management. The implementation of such technologies requires substantial investment in both hardware and skilled personnel capable of using and maintaining the systems, which should be accounted for in any business case. Furthermore, effective collaboration between transportation companies and stakeholders is needed to share data and improve the accuracy and efficiency of these planning tools.

Investing in improved infrastructure is fundamental. This includes constructing more efficient rail lines, upgrading port facilities to accommodate larger and more environmentally friendly vessels, and developing better road networks that support electric and hybrid vehicles. Such upgrades need to be carefully planned to minimize environmental disruptions during construction and ensure long-term sustainability. The environmental impact assessments must be thorough, public participation encouraged, and community needs taken into account throughout the project's lifecycle.

Promoting modal shift towards more environmentally friendly modes is a significant step towards sustainability. Encouraging the shift of freight from road to rail or water transport, for example, can significantly reduce emissions per unit transported. Government incentives, such as tax breaks or subsidies for the use of more sustainable modes, can encourage this shift. However, a comprehensive approach is needed, addressing factors like cost-competitiveness and the overall efficiency of the different transport modes. This requires detailed analysis and a potential reassessment of pricing structures to make sustainable choices economically viable.

Sustainable intermodal transport also requires a broader societal approach. This involves educating consumers, businesses, and governments on the importance of environmental responsibility and the benefits of sustainable transportation. Promoting eco-friendly purchasing decisions, supporting businesses that prioritize sustainability, and advocating for policies that incentivize environmentally conscious practices are all part of this broader movement. It means integrating sustainability into every aspect of the decision-making process, from the initial design and planning to the final disposal or recycling of materials.

Furthermore, implementing effective carbon offsetting programs is a necessary part of the solution. While reducing emissions remains the primary focus, investments in projects that absorb or remove carbon dioxide from the atmosphere can help compensate for unavoidable emissions. These programs, however, must be carefully monitored to ensure their effectiveness and transparency, preventing any "greenwashing" efforts that might undermine genuine progress towards sustainability. Careful selection of certified projects and transparent reporting are paramount.

Ultimately, achieving sustainable solutions in intermodal transport requires a collaborative effort involving governments, industry stakeholders, and consumers. Only through a coordinated approach involving technological advancements, infrastructural improvements, policy changes, and a fundamental shift towards environmental responsibility can we create a truly sustainable and efficient global transportation network. This needs to go beyond short-term fixes and involve long-term strategic planning, with mechanisms for continuous monitoring, evaluation, and improvement of sustainability strategies. The journey towards sustainable intermodal transport is a continuous process of adaptation and improvement, necessitating ongoing commitment and collaboration among all stakeholders.

7.2 Global Trade Impacts

Global trade relies heavily on efficient and interconnected transportation systems. The seamless movement of goods across borders fuels economic growth, supports international cooperation, and ultimately shapes the global economy. Disruptions, whether due to natural disasters, geopolitical instability, or logistical bottlenecks, can have cascading effects, impacting everything from consumer prices to national security. Understanding these impacts is paramount for policymakers, businesses, and anyone involved in international commerce.

Consider the impact of a major port closure. A significant disruption at a major container hub, like Singapore or Rotterdam, can trigger a chain reaction. Delays ripple through the supply chain, causing shortages of goods, increased prices, and potentially even factory closures. The ripple effect isn't limited to the immediate vicinity; it cascades globally, affecting businesses and consumers in far-flung markets. This demonstrates the interconnected nature of global trade and the vulnerability of systems dependent on efficient transportation.

The growth of e-commerce has dramatically increased the demand for rapid and reliable delivery. Consumers now expect goods to arrive quickly, regardless of origin. This pressure has fueled innovations in logistics, including the greater use of air freight and faster, more efficient ground transport. However, this heightened demand also puts additional strain on existing infrastructure and creates new challenges for managing the flow of goods. Increased reliance on air freight, for example, raises concerns about carbon emissions and environmental sustainability.

Geopolitical events can significantly disrupt global trade flows. Trade wars, sanctions, and political instability can lead to border closures, trade restrictions, and disruptions to shipping routes. These disruptions can have devastating economic consequences, not only for the countries directly involved but also for the global economy as a whole. Supply chains become fragile, and businesses are forced to adapt quickly or risk financial losses. Diversification of supply chains, a strategy once considered optional, now becomes a necessity for business continuity.

Technological advancements, while offering tremendous opportunities for improvement, also present their own set of challenges. The implementation of new technologies, such as automated container handling systems or advanced tracking software, requires significant investment and expertise. Furthermore, cybersecurity threats pose a growing risk to the security and integrity of global trade systems. Data breaches and cyberattacks can disrupt operations, compromise sensitive information, and even lead to significant financial losses. Therefore, robust cybersecurity measures are crucial for maintaining the integrity of modern intermodal networks.

Furthermore, the increasing complexity of global supply chains requires better coordination and communication among all stakeholders. This includes not only transportation providers but also manufacturers, importers, exporters, and customs officials. Effective communication and collaboration are vital for preventing bottlenecks, avoiding delays, and ensuring that goods reach their destinations on time and in good condition. The ability to track goods in real-time, predict potential delays, and proactively address disruptions becomes increasingly important in navigating the complexities of modern global trade.

The economic implications of disruptions to global trade are profound. Increased transportation costs can lead to higher prices for consumers. Businesses may face reduced profitability due to delays and increased logistical challenges. Ultimately, the disruption of global trade can negatively impact economic growth and social stability. Effective management of these risks is therefore a critical component of ensuring the smooth functioning of the global economy.

In conclusion, the impacts of global trade extend far beyond the simple movement of goods. They have profound consequences for economies, societies, and the environment. The challenges and opportunities presented by global trade require a holistic approach, integrating technological advancements, robust security measures, and effective collaboration among all stakeholders. A well-functioning intermodal transport system is critical to mitigating risks and maximizing the benefits of global trade. Therefore, a forward-looking and adaptive strategy is essential for navigating the complexities and harnessing the potential of interconnected global commerce in the years to come. The development and implementation of such strategies are paramount for ensuring the stability and prosperity of the global economy.

7.3 Emerging Technologies

Autonomous vehicles are rapidly changing the landscape of intermodal transport. Their potential to streamline logistics and reduce reliance on human drivers is enormous. Imagine a future where self-driving trucks seamlessly transfer containers between rail yards and distribution centers, operating 24/7 with minimal downtime. This technology promises increased efficiency, cost savings through reduced labor costs, and potentially improved safety due to reduced human error. However, significant challenges remain, including the need for robust infrastructure to support autonomous navigation, addressing cybersecurity vulnerabilities, and navigating the complex regulatory landscape surrounding autonomous vehicle operation. The successful integration of autonomous vehicles will require considerable investment in infrastructure, technological development, and regulatory frameworks to ensure safe and reliable operation.

The integration of drone technology offers another compelling avenue for innovation in intermodal transport, particularly for last-mile delivery. Drones can access areas inaccessible to traditional vehicles, such as remote locations or congested urban environments. This capability significantly reduces delivery times, particularly for time-sensitive goods, and minimizes the environmental impact of last-mile delivery, which currently relies heavily on road transport. However, challenges such as payload limitations, battery life, air traffic management regulations, and public acceptance need to be overcome. The future of drone delivery is likely to involve a hybrid approach, integrating drones with other transportation modes for optimal efficiency. Addressing these limitations through further technological development and collaborative regulatory frameworks is crucial for realizing the full potential of drones in intermodal logistics.

3D printing is emerging as a disruptive technology with significant implications for intermodal transport. While not directly involved in the movement of goods, 3D printing has the potential to revolutionize the manufacturing and repair of components used across the various modes of transport. Imagine on-site repair of damaged containers or the rapid production of specialized parts for railcars or ships, minimizing downtime and reducing the need to transport replacement parts across long distances. This technology reduces lead times, lowers inventory costs, and promotes sustainability by minimizing transportation-related emissions. The integration of 3D printing into intermodal logistics requires careful consideration of material selection, printing speed, scalability, and the skilled workforce required to operate and maintain 3D printing equipment.

Advanced sensor technologies are playing a transformative role in improving the safety and efficiency of intermodal transportation. Real-time monitoring of cargo conditions, using sensors that track temperature, humidity, and other vital parameters, ensures the integrity of goods throughout their journey. This reduces losses from spoilage and damage, benefiting both shippers and carriers. Similarly, sensors embedded in infrastructure, such as rail lines or ports, can detect potential hazards, such as track defects or structural weakness, preventing accidents and minimizing downtime. The data collected by these sensors provides valuable insights into operational efficiency, allowing for proactive maintenance and optimization of transportation networks. However, the effective use of sensor data requires advanced data analytics capabilities to process the vast amounts of information generated, and robust cybersecurity measures to protect the integrity of this information.

The application of artificial intelligence (AI) and machine learning (ML) is revolutionizing various aspects of intermodal transport. These technologies can analyze vast datasets to optimize routing, predict delays, and improve resource allocation. AI-powered systems can analyze weather patterns, traffic conditions, and other factors to dynamically adjust schedules and improve the overall efficiency of the transportation network. Predictive maintenance using AI can identify potential equipment failures before they occur, reducing downtime and ensuring the smooth flow of goods. However, the integration of AI and ML into intermodal systems requires careful consideration of data security, algorithm bias, and the need for human oversight to ensure responsible and ethical deployment. The development of robust and explainable AI models is crucial for building trust and ensuring the successful implementation of these technologies.

The development and application of these emerging technologies will continue to reshape the intermodal landscape, creating more efficient, resilient, and sustainable transportation systems. The successful integration of these technologies requires collaboration between industry stakeholders, policymakers, and technology providers to address the associated challenges and to ensure responsible and beneficial deployment. This collaborative effort is essential to unlock the full potential of these innovations and to pave the way for a future where intermodal transportation is both efficient and sustainable. The future of intermodal transport depends on our ability to embrace and effectively integrate these advanced technologies.

www.ingramcontent.com/pod-product-compliance
Lightning Source LLC
LaVergne TN
LVHW022013200926
848481LV00007B/2363